SpringerBriefs in Education

We are delighted to announce SpringerBriefs in Education, an innovative product type that combines elements of both journals and books. Briefs present concise summaries of cutting-edge research and practical applications in education. Featuring compact volumes of 50 to 125 pages, the SpringerBriefs in Education allow authors to present their ideas and readers to absorb them with a minimal time investment. Briefs are published as part of Springer's eBook Collection. In addition, Briefs are available for individual print and electronic purchase.

SpringerBriefs in Education cover a broad range of educational fields such as: Science Education, Higher Education, Educational Psychology, Assessment & Evaluation, Language Education, Mathematics Education, Educational Technology, Medical Education and Educational Policy.

SpringerBriefs typically offer an outlet for:

- An introduction to a (sub)field in education summarizing and giving an overview of theories, issues, core concepts and/or key literature in a particular field
- A timely report of state-of-the art analytical techniques and instruments in the field of educational research
- A presentation of core educational concepts
- An overview of a testing and evaluation method
- A snapshot of a hot or emerging topic or policy change
- An in-depth case study
- A literature review
- A report/review study of a survey
- An elaborated thesis

Both solicited and unsolicited manuscripts are considered for publication in the SpringerBriefs in Education series. Potential authors are warmly invited to complete and submit the Briefs Author Proposal form. All projects will be submitted to editorial review by editorial advisors.

SpringerBriefs are characterized by expedited production schedules with the aim for publication 8 to 12 weeks after acceptance and fast, global electronic dissemination through our online platform SpringerLink. The standard concise author contracts guarantee that:

- an individual ISBN is assigned to each manuscript
- each manuscript is copyrighted in the name of the author
- the author retains the right to post the pre-publication version on his/her website or that of his/her institution

Claude Müller

Digital Learning Design

Designing Effective Online and Blended Learning

 Springer

Claude Müller
Zurich University of Applied Science
Winterthur, Switzerland

ISSN 2211-1921   ISSN 2211-193X   (electronic)
SpringerBriefs in Education
ISBN 978-3-031-89047-5   ISBN 978-3-031-89045-1   (eBook)
https://doi.org/10.1007/978-3-031-89045-1

This book is an open access publication.

This Springer imprint is published by the registered company Springer Nature Switzerland AG
The registered company address is: Gewerbestrasse 11, 6330 Cham, Switzerland

If disposing of this product, please recycle the paper.

Preface

The shift from traditional teaching to digital learning presents a significant challenge for many educators. Navigating the complexities of digital course designs can often lead to suboptimal learning experiences that fail to engage learners effectively. *Digital Learning Design: Designing Effective Online and Blended Learning* aims to address this challenge by offering a comprehensive guide that combines cognitive science principles with practical design strategies. This book equips learning professionals—including trainers, teachers, and digital learning experts—with the knowledge and tools needed to create impactful and attractive online and blended learning environments to foster both meaningful engagement and knowledge retention in today's digital landscape.

This book is designed to be read sequentially, from Chaps. 1 to 9. Each chapter builds on the previous one, offering conceptual foundations, evidence-based insights, and practical design tips. The structure mirrors the digital learning design process, allowing readers to grasp both theory and application simultaneously. More experienced designers may choose to read individual chapters based on their specific interests or needs. Practitioners looking for hands-on advice can focus primarily on the practical tips included throughout the book. Additionally, users of the myScripting tool will find in-depth educational concepts and design considerations linked to the core functionalities of the tool.

The main topics covered in this book are as follows:

1. **Introduction:** Explore the opportunities and challenges of digital learning, examining both its potential benefits and the obstacles you may encounter as you integrate technology into your teaching. Gain clarity on the various terms used to describe learning with media and technology, ensuring you have a solid understanding of digital education terminology.
2. **Principles of Cognitive Science:** Dive into the foundational principles of cognitive science to understand how learners process and retain information. By grasping these underlying mechanisms, you can enhance your teaching, ultimately improving both the effectiveness of your digital learning environment and your learners' motivation.

3. **Design Models:** Discover the key characteristics of teaching and learning situations and see how they are interconnected and influence one another. This chapter outlines the essential steps and phases for developing digital learning environments and highlights proven approaches that lead to successful outcomes. You will also discover overarching design principles that support the creation of meaningful and engaging digital learning experiences.
4. **Context Analysis:** Equip yourself with a systematic approach to context analysis, starting with needs assessment and learner analysis. Determine whether there is a genuine need for training, identify your target group, and explore their characteristics in detail. Explore key competencies to develop and learning outcomes to achieve while gaining an understanding of the resources available for programme development to ensure a thorough understanding of the context before designing your digital learning experience.
5. **Learning Organization:** Explore various forms of learning organization, focusing on how to structure time and space in digital learning. Analyse the advantages and disadvantages of asynchronous versus synchronous and on-site versus online learning, and gain insights into when each approach is most effective. You will also examine the defining characteristics of blended and online learning formats, giving you a clear understanding of their unique features and how to enhance the learning experience.
6. **Content Structuring:** Discover techniques for creating clear and engaging content that aligns with your course objectives. This chapter explains how learning outcomes help define the appropriate level of abstraction and complexity. Learn to prioritize, simplify, segment, and sequence content within the constraints of working memory to ensure effective absorption by learners.
7. **Teaching Strategies:** Take an in-depth look at a variety of teaching strategies that shape the educational design of effective learning environments. You will discover the key characteristics of different approaches, focusing on comparing direct instruction and inquiry learning. Learn how to determine which strategy best fits a specific educational context and uncover the essential criteria for making informed choices.
8. **Learning Activities:** Discover methods for designing learning activities that activate prior knowledge and foster interaction. You will learn practical guidelines for creating engaging activities focused on activation, interaction, and assessment, along with tips for selecting appropriate learning media. This approach ensures that your activities not only replicate traditional methods but also extend and transform your teaching practices in digital environments.
9. **Reflection:** Learn how to systematically review the design process and assess whether you have met the key steps and conditions for creating effective learning experiences. This chapter emphasizes the importance of reflection in ensuring that your learning environment meets educational goals and enhances learner engagement and success.

The book has been developed over the last few years as part of my work at my university, alongside my teaching and research activities. This content is an integral part of a study programme called digital learning, which is delivered in a blended learning format. We have also been transforming several study programmes into a blended learning format and have been producing online courses in collaboration with company partners or for a massive open online courses (MOOC) platform. All these projects have enabled us to test the concepts and practical strategies outlined in this manual in real blended and online learning settings over the course of a decade. Early versions of this manual were used as prototypes by the initiatives, and I am deeply grateful to the project partners and participants who provided valuable feedback on the manuscript.

Over the past few years, we have also developed the myScripting educational design tool to help educators design effective and engaging digital learning environments. This handbook highlights the core concepts and processes that underlie the key features of myScripting, and the implementation in myScripting is shown at the end of each chapter. Together, the tool and this manual should give you the ability to create digital learning environments that are not only effective but also engaging and efficient. In addition to the information in this guide, you will find tutorials and instructional videos at https://myscripting.zhaw.ch.

Winterthur, Switzerland   Claude Müller

Contents

Chapter 1
Introduction

As the digitalization of society continues, there is a growing demand for digital learning in a blended or online learning format that will better meet the varying needs of learners while also making learning more effective, attractive, and efficient. At the same time, there are concerns about the risks of social isolation, especially in online learning.

This chapter explores the following key questions: What are the opportunities and challenges of digital learning? What terms are used to describe learning with media and technology?

1.1 Potential and Challenges in Digital Learning

Digital learning offers numerous opportunities to enhance and support educational processes. First, it makes learning independent of time and space, giving learners the flexibility to adapt their education to their individual needs and life contexts. Multimedia elements can also enrich learning by illustrating content more effectively and immersing learners in authentic contexts. For instance, virtual reality (VR) can provide safe and controlled environments for realistic exercises, such as surgical or flight simulations, which would be too dangerous or impractical in the real world. Additionally, interactive and adaptive tools allow for personalized learning. For example, language learning tools can continuously assess learners' competencies and adjust content accordingly, while communication tools support collaborative learning in increasingly remote settings. This potential demonstrates that digital learning can significantly expand and transform traditional educational methods (see, for example, the PICRAT model of Kimmons et al. 2020).

Moreover, digital tools can support teachers by facilitating planning, organizing, delivering, and evaluating lessons, relieving them of routine tasks and allowing more time for valuable pedagogical interactions, such as providing feedback and

© The Author(s) 2025
C. Müller, *Digital Learning Design*, SpringerBriefs in Education,
https://doi.org/10.1007/978-3-031-89045-1_1

supporting learners' progress. Tools such as learning management systems (LMSs) can provide valuable services in this respect.

However, digital learning also presents challenges. For example, the flexibility it offers can result in less direct and often delayed interactions, making it harder to identify and address comprehension issues in real time. Additionally, online learning contains many distractions. As a result, special attention must be paid to motivational factors when designing digital learning environments, as social cohesion between learners—and between learners and teachers—tends to be weaker due to physical distance, increasing the risk of social isolation.

Research has indicated that blended and online learning formats often achieve learning outcomes comparable to, or even better than, traditional formats (Bernard et al. 2014; Means et al. 2013; Müller and Mildenberger 2021). However, these results can be highly variable and depend largely on the quality of the implementation (Müller et al. 2023). The effectiveness of digital learning, therefore, relies less on the media or tools used and more on the quality of the educational design in the specific learning context (Clark 1994; Reigeluth and Honebein 2023). Consequently, when designing digital learning environments, the focus should not be on the latest technology or tools but on how these tools can best support and enhance the learning process.

Many teachers have limited experience or expertise in systematically developing digital learning environments. As a result, they often transfer existing teaching concepts to digital formats, such as using digital flashcards instead of paper or holding group work in breakout rooms rather than in physical rooms. This approach, as seen during the pandemic, is not necessarily ineffective—breakout rooms, for example, were particularly successful in facilitating social interaction. However, simply transferring existing methods to digital tools does not fully exploit the potential of digital learning, nor does it significantly improve or transform pedagogical practice.

The goal of this handbook is to introduce teachers to evidence-based approaches for designing digital learning and to support them in using these approaches. "Evidence-based" means that the processes and principles presented are grounded in current research on teaching and learning with media and technology.

The myScripting tool supports the educational design process electronically, and references to the tool are included throughout this handbook. Together, the handbook and tool aim to enhance teachers' competence in designing effective digital learning environments. With reference to the widely used *TPACK competency model* (Mishra and Koehler 2006), this handbook addresses the intersections among pedagogy, technology, and content, focusing on how to design learning environments that integrate these elements effectively (see Fig. 1.1). The focus is on the educational design of digital learning rather than on technical principles or tools. The central question is this: How can technology be used to support and enhance teaching and learning?

In the first section, key terms and concepts related to digital learning are clarified, and essential principles of cognitive science, such as human information processing and motivation, are introduced. Designing digital learning environments often requires collaboration between specialists (e.g., subject matter experts [SMEs],

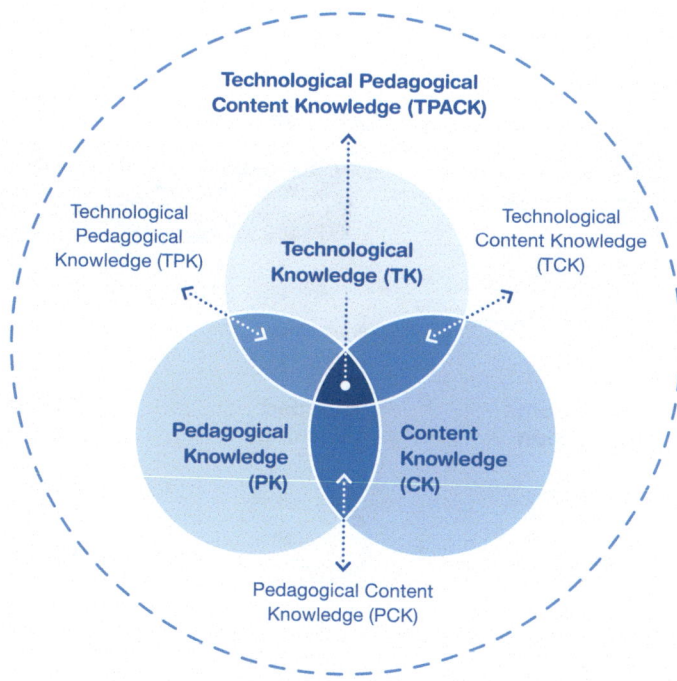

Fig. 1.1 TPACK competency model for digital learning design (Mishra and Koehler 2006)

educational designers, and media producers). A shared understanding of pedagogical terms and principles can facilitate this collaborative process. In later chapters (see Chaps. 3–8), a systematic design process for digital learning is presented, outlining the development steps, design options, and critical design decisions.

1.2 Key Terms and Concepts

Many terms are used in the context of teaching and learning with media and technology, but their meanings often vary across fields. These terms are generally characterized by the technologies involved and the instructional methods applied. Figure 1.2 provides an overview of common terms, illustrating their use along a proximity–distance continuum, with fluid transitions between concepts.

Digital learning is a broad term, encompassing all teaching and learning processes that involve *information and communication technology (ICT)*. This includes technologies such as virtual or augmented realities (VR/AR), simulations, and, more recently, data analysis or artificial intelligence (AI) for purposes such as learning support through learning analytics. Digital learning often refers to these newer possibilities, while *e-learning* is used when learning primarily involves computers over networks, such as the internet. LMSs serve as comprehensive platforms that

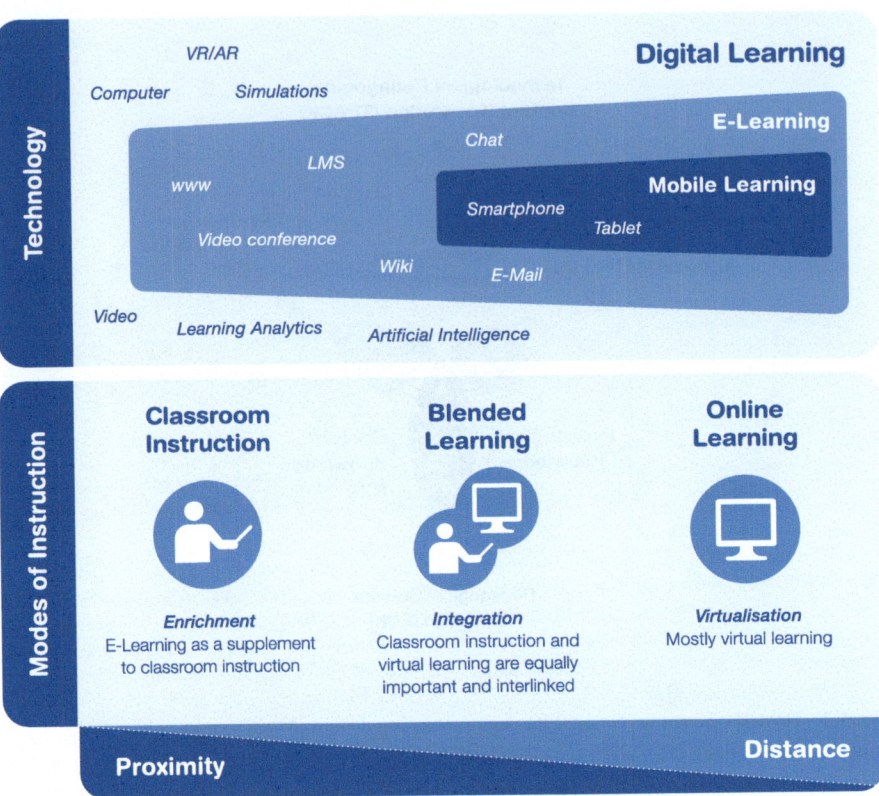

Fig. 1.2 Key terms and concepts used in teaching and learning with media and technologies

often integrate tools for video conferencing and other interaction and communication tools, such as wikis, chats, or forums. The term "e-learning" is somewhat older and is gradually being replaced by the newer term "digital learning." *Mobile learning* is a more narrowly defined term focused on learning through portable devices, such as smartphones, laptops, or tablets. In this handbook, the term "digital learning" primarily refers to teaching and learning using media and technology.

Digital learning can be used to varying degrees in face-to-face instruction, guided self-study, and autonomous self-study. From a pedagogical perspective, three main concepts are relevant: *enrichment*, *integration*, and *virtualisation* (see Fig. 1.2).

In the *enrichment* approach, digital tools complement traditional classroom teaching. For example, learners can engage with interactive resources, such as animations and simulations, or explore dynamic content, such as instructional videos, accessible online. Additionally, communication platforms—including forums and chat functions—can facilitate interactions and provide support for learners. In the enrichment approach, digital elements are offered alongside classroom teaching. For instance, learners may access interactive resources (e.g., animations, simulations) or dynamic content (e.g., learning videos) electronically. Communication tools, such as forums and chats, can also support the course.

In the *integration approach*—also known as *blended learning*—face-to-face and virtual learning phases are combined into an interconnected learning format. Traditionally, face-to-face instruction occurs synchronously (in real time) with an exchange between the teacher and learners. Although this is often understood as on-site classes, tools such as Zoom have made virtual face-to-face interactions more common (also referred to as *virtual blended learning*). *Hybrid learning* was previously synonymous with blended learning, but during the pandemic, it increasingly referred to sessions in which learners could choose between on-site and online participation. In addition to synchronous phases, blended learning includes asynchronous components, in which learners engage in self-paced learning in a virtual environment. Examples of these virtual phases include interactive exercises (e.g., tests with automatic feedback), collaborative online group work, and tutorial support through online communication during self-study.

In the *virtualisation concept*, the learning environment consists primarily or entirely of virtual components, known as *online learning* or *distance learning*. This model focuses on bridging the physical distance between teachers and learners using information and communication technologies. While distance learning has existed for a long time (previously through physical materials), today, it is largely implemented through digital technologies.

In the concrete design of a digital learning environment, the individual components are referred to as *learning activities*. Examples include reading an article, contributing to an interactive forum, or taking an online test. A *learning assignment* can initiate and structure one or more learning activities (see Fig. 1.3). Learning assignments should guide learners through the digital learning offering and support their learning process.

It is also essential that these activities take place in different *social forms of learning*. In addition to individual work, learners can engage in partner work, small

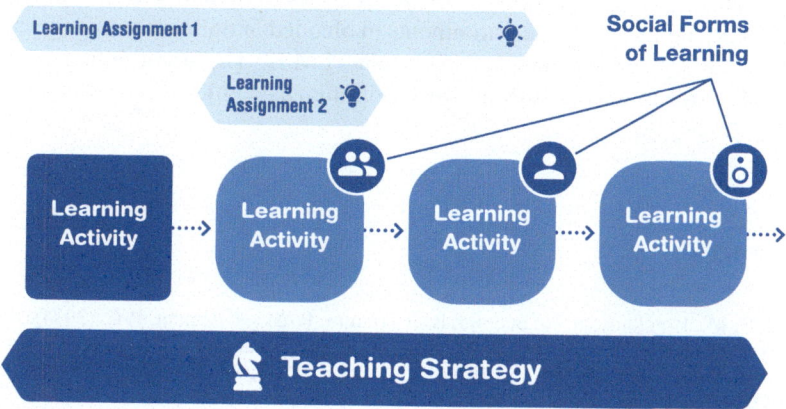

Fig. 1.3 Key terms and concepts used in educational design

group collaboration (3–5 people), large group discussions (6–12 people), or plenary sessions (over 12 people).

A typical sequence of learning activities is called a *teaching strategy*. Well-known strategies include direct instruction and problem-based learning. Depending on the educational context, one strategy may be more appropriate than another when designing a specific digital learning environment. Different teaching strategies are discussed in more detail in Chap. 7.

 **Tips for Digital Learning**

- **Transform education:** Use digital tools and media not just to replace your own teaching into virtual spaces but to enhance and transform the learning experience.
- **Focus on pedagogy:** Prioritize educational design over the choice of tools or media when developing digital learning environments.
- **Define key terms:** Establish a common understanding by clarifying key digital learning terms with both teachers and learners.

myScripting   **Introduction to myScripting**

myScripting is a web-based visual educational design tool that supports teachers in the conception, implementation, and continuous development of learning offerings. It allows users to map and plan teaching sequences along a temporal axis using a flowchart format. Each learning phase and activity can be enriched with metadata (e.g., workload, tools, assessments, or learning assignments), and the educational design, along with the stored metadata, can be exported in various formats. myScripting is particularly well suited for designing digital learning environments in blended or online learning formats, as its activity set is tailored to electronic LMSs. The myScripting tool is available free of charge at [https://myscripting.zhaw.ch].

References

Bernard, R. M., Borokhovski, E., Schmid, R. F., Tamim, R. M., & Abrami, P. C. (2014). A meta-analysis of blended learning and technology use in higher education: From the general to the applied. *Journal of Computing in Higher Education, 26*(1), 87–122. https://doi.org/10.1007/s12528-013-9077-3

Clark, R. E. (1994). Media will never influence learning. *Educational technology research and development, 42*(2), 21–29. https://doi.org/10.1007/BF02299088

Kimmons, R., Graham, C. R., & West, R. E. (2020). The PICRAT model for technology integration in teacher preparation. *Contemporary Issues in Technology and Teacher Education, 20*(1), 176–198.

Means, B., Toyama, Y., Murphy, R., & Baki, M. (2013). The effectiveness of online and blended learning: A meta-analysis of the empirical literature. *Teachers College Record, 115*(3), 1–47. http://www.tcrecord.org/Content.asp?ContentId=16882

Mishra, P., & Koehler, M. J. (2006). Technological pedagogical content knowledge: A framework for teacher knowledge. *Teachers college record, 108*(6), 1017–1054. https://doi.org/10.1111/j.1467-9620.2006.00684.x

Müller, C., & Mildenberger, T. (2021). Facilitating flexible learning by replacing classroom time with an online learning environment: A systematic review of blended learning in higher education neu. *Educational Research Review, 34*, 100394. https://doi.org/10.1016/j.edurev.2021.100394

Müller, C., Mildenberger, T., & Steingruber, D. (2023). Learning effectiveness of a flexible learning study programme in a blended learning design: Why are some courses more effective than others? *International Journal of Educational Technology in Higher Education, 20*(1), 10. https://doi.org/10.1186/s41239-022-00379-x

Reigeluth, C. M., & Honebein, P. C. (2023). Will instructional methods and media ever live in unconfounded harmony? Generating useful media research via the instructional theory framework. *Educational technology research and development.* https://doi.org/10.1007/s11423-023-10253-w

Chapter 2
Principles of Cognitive Science

A deep understanding of cognitive and motivational learning processes is crucial for designing effective digital learning environments. By aligning the structure and processes of learning with how the brain functions, learners' ability to build knowledge can be significantly enhanced, leading to improved learning outcomes. Equally important are motivational processes. Understanding the factors that drive motivation enables the design of digital experiences that capture and sustain learners' interests while fostering engagement.

This chapter explores the following key questions: How does human learning work? What are the implications of information processing for designing digital learning environments? How can learners be motivated to actively use digital environments and engage in the learning process?

2.1 Information Processing

The cognitive processes involved in acquiring, processing, storing, and retrieving information are fundamental to designing effective learning environments. These processes rely on memory, which consists of three key components: *sensory memory, working memory*, and *long-term memory* (Atkinson and Shiffrin 1968; see Fig. 2.1). Understanding how these memory systems interact is crucial for optimizing educational design.

Sensory memory rapidly captures large amounts of visual and auditory information but retains it only briefly (milliseconds). When attention is focused on this information, it transfers to working memory, where processing occurs (e.g., language comprehension or problem-solving) and is reorganized into manageable units of information, called chunks, for efficient storage in *long-term memory*.

Working memory includes a *central executive* that manages attention, along with two subsystems: the *visuo-spatial sketchpad* for visual information and the *phonological loop* for verbal information (Baddeley 1986). Information is elaborated and

© The Author(s) 2025
C. Müller, *Digital Learning Design*, SpringerBriefs in Education,
https://doi.org/10.1007/978-3-031-89045-1_2

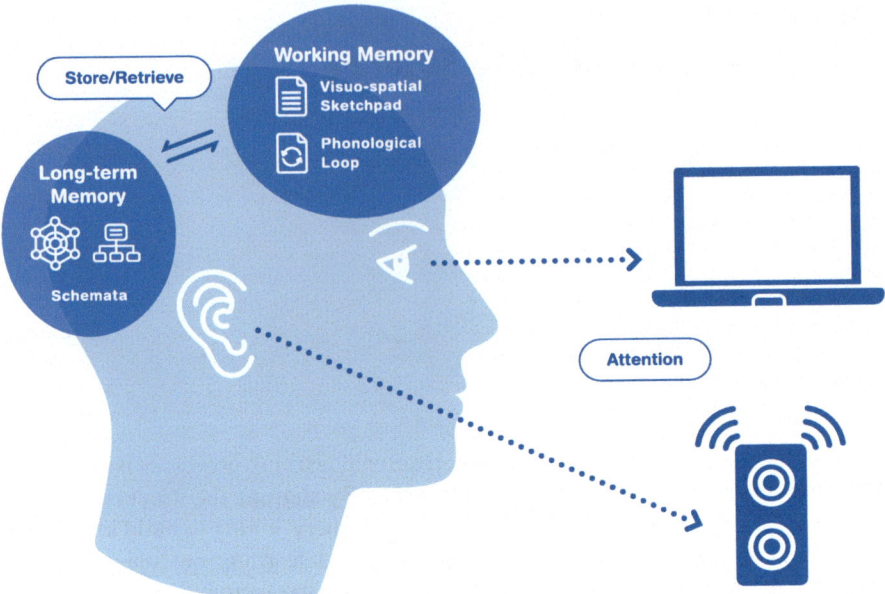

Fig. 2.1 Information processing and knowledge building

stored in both visual and linguistic formats (Paivio 1986). Learning is enhanced when information is presented simultaneously as verbal content and images, a principle known as the *multimedia principle* (see also Sect. 8.1).

Long-term memory is the core structure of human cognition and plays a critical role in overall cognitive performance. It stores vast amounts of information indefinitely, but effective retrieval depends on organized pathways or "maps." Knowledge in long-term memory is structured as schemata—categories of information organized by their use. As more schemata are developed and become more complex, they integrate elements from simpler schemata. This allows vast amounts of information to be consolidated into a single schema, which working memory treats as one unit, facilitating the efficient execution of complex tasks.

Working memory, however, is limited in both time and capacity. If information is not revisited quickly, it may be lost and must be reconstructed from long-term memory. Additionally, working memory can handle only a limited number of information units (often referred to as "seven plus/minus two") at any given time (Miller 1956). *Cognitive load* theory highlights the constraints of working memory and seeks to reduce unnecessary load to facilitate long-term memory development (Sweller et al. 1998).

Cognitive load consists of intrinsic and extrinsic components. *Intrinsic load* refers to the complexity and interactivity of the learning material, which varies depending on the learner's expertise. Experienced learners who can retrieve information from long-term memory face a lower intrinsic load than novices. Additionally, the germane load, once considered an independent component, is now seen as part of the intrinsic load relating to the mental effort required to process and understand the material (Sweller et al. 2019). *Extrinsic load*, on the other hand, is influenced by the design of the learning environment. Poor design—such as excessive cross-references, redundant material, or unclear layouts—can unnecessarily increase extrinsic load and hinder learning. Minimizing extrinsic load is particularly important for learners with less prior knowledge.

To optimize learning, the sum of intrinsic and extrinsic loads should not exceed the capacity of the working memory. If the load is too high, content can be prioritized, simplified in terms of educational adaptation, divided into smaller segments (segmentation), and sequenced in meaningful order. Chapter 6 (Content Structuring) discusses strategies for reducing cognitive load through segmentation and sequencing.

In summary, when designing learning environments, the aim should be to optimize cognitive resources by reducing extrinsic load and managing intrinsic load in alignment with learners' expertise. This ensures that the learning process remains efficient and effective.

These design principles are particularly important for novices, whose working memory is easily overwhelmed. However, for advanced learners, research shows that learning environments optimized according to these principles may offer little benefit or even hinder learning—this is known as the *expertise reversal effect* (Kalyuga et al. 2003). This occurs because experienced learners must adapt their existing cognitive schemata to accommodate redundant information, which places more strain on their working memory than it does for novices. Therefore, the degree of structuring and support, named *scaffolding*, in educational design should be tailored to the learner's level of expertise, as illustrated in Fig. 2.2, which shows how scaffolding is adapted for different learning tasks.

In a *worked example*, the full solution and solution steps are provided, allowing learners to focus their working memory on understanding the underlying principles. In *completion tasks*, learners are given partial steps and a partial solution, which

Fig. 2.2 Scaffolding of the learning task

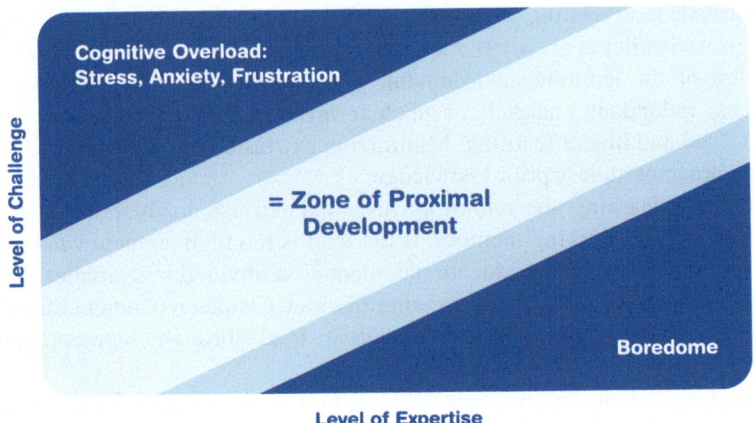

Fig. 2.3 Zone of proximal development (based on Vygotsky 1978)

they must use to solve the task themselves. *Problem-solving tasks* provide even less support, requiring learners to select the appropriate strategy and solve the problem independently. These tasks are suited to advanced learners with substantial knowledge and motivation.

The complexity of learning tasks can be adjusted based on the degree of scaffolding. Ideally, there should be a balance between the level of challenge of the task and the learner's level of expertise, aligning with the *zone of proximal development* (Fig. 2.3). This concept, developed by Vygotsky (1978), suggests that learners achieve the best results when presented with tasks that are slightly beyond their current abilities but achievable with guidance from teachers, peers, or learning materials. *Adaptive learning* systems can increasingly diagnose a learner's competency and assign tasks that offer the right level of challenge within the zone of proximal development.

When adapting learning environments to learners, one frequently debated topic is adaptation to learners' preferred learning styles, such as visual or auditory preferences. Learning style theory was debunked as a myth very early on, but it continues to crop up in discussions about the design of learning environments (Kirschner 2017). While learners may have personal preferences, there is no evidence that tailoring instruction to these preferences improves learning outcomes (Neelen and Kirschner 2020). Similarly, the notion of differentiating instruction based on generational traits, such as those associated with millennials or Generation Z, is also a myth without empirical support.

2.2 Learning Motivation and Engagement

The cognitive psychology theories outlined above focus primarily on the cognitive aspects of information processing. A key conclusion from these theories is the principle that "less is more"—information should be reduced and simplified for an effective learning process. However, these theories often overlook the motivational aspects of learning. For example, images not only serve to visualize content but also play a motivational role in the learning process (Schneider et al. 2016). They can draw attention to learning materials. Learners need to be prepared to actively use cognitive resources and engage in the learning process (Kahu 2013). This is particularly critical in asynchronous online learning environments, where distractions are abundant and learners must focus on the task and activate their motivation. Therefore, recent models of learning in digital environments have integrated social and affective theories alongside cognitive ones (Moreno 2006; Park et al. 2014; Schneider et al. 2022).

Learner Motivation

Explaining learner motivation is complex, with various theories addressing the subject (Cook and Artino 2016). Factors such as the learning context, individual learner characteristics, and the design of the learning environment play critical roles in determining whether learners will engage with the material and perform learning actions.

Expectancy-value theory (Eccles et al. 1983) provides a psychological framework to explain individual motivation and decision-making, especially in educational tasks such as completing assignments or achieving specific outcomes (see also Fig. 2.4). The theory suggests that individuals will be motivated to engage in a task if they have an *expectation of success* and if they perceive *value* in the task.

In digital learning, as in other forms of learning, the *expectancy of success* goes beyond a general perception of one's own competence. This reflects a

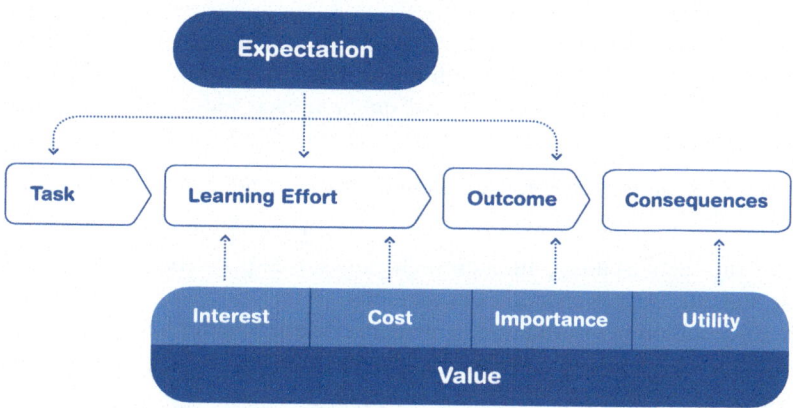

Fig. 2.4 Expectancy-value theory (based on Eccles et al. 1983)

forward-looking belief that one can successfully complete a task. If learners doubt their ability to succeed, they will lack the motivation to begin. The expectancy of success is influenced by self-concept and perceived task difficulty. Self-concept refers to the learner's belief in their abilities (academic, social, or other skills), while task difficulty is based on how challenging the learner perceives the task to be. Empirical studies show that expectancy of success can predict engagement in learning activities and learning outcomes, such as grades or test scores (Cook and Artino 2016).

However, expectancy alone is not enough. According to expectancy-value theory, learners must also perceive an immediate or future personal benefit or *value* from the task. Similar to expectancy, task value is subjective and depends on four factors (Eccles and Wigfield 2020):

- *Interest:* The learner's intrinsic interest in a topic or activity.
- *Importance:* The task's alignment with the learner's personal goals or self-concept.
- *Utility:* The perceived usefulness of the task for practical reasons or for achieving future goals.
- *Cost:* The effort required to complete the task, including opportunity costs, such as time and energy.

Motivation is influenced by *perception*, not necessarily by reality. Factors such as affective memories (reactions and emotions related to past experiences) and social influences (such as parents, teachers, peers, and professional values) shape perceptions of expectations of success and the value of a task. This also means that what motivates one person may not necessarily motivate another in the same way.

Designing Motivating Digital Learning Environments

When designing digital learning environments, it is essential to create incentives that make learning activities engaging, appealing, and personally valuable to learners. At the same time, the task's difficulty should be aligned with the learner's expertise (zone of proximal development; see Fig. 2.3) to ensure that learners feel capable of success.

The *ARCS model* (Keller 1983) provides practical strategies based on motivational psychology to foster motivation in learning environments. ARCS stands for *attention, relevance, confidence,* and *satisfaction*, each addressed through the following measures.

Attention: How do you capture and maintain learners' attention without distracting them from the learning content?

- Encourage active participation through games, role-plays, or simulations.
- Use humour, such as anecdotes or cartoons.
- Present conflicting information to challenge learners' prior knowledge.
- Diversify teaching with different media and strategies.
- Present current real-world examples and problems.

Relevance: Why should learners care about this content?

- Relate content to learners' existing skills and experiences.
- Emphasize the usefulness of the content in current or future professional or private situations.
- Use modelling to demonstrate how content applies and what learners can do with it.
- Offer choices to let learners engage in different ways and showcase their competencies.

Confidence: How do you make learners feel they can succeed?

- Clearly communicate learning requirements, objectives, and assessments so that learners can assess their chances of success and plan the necessary learning steps.
- Provide a sense of achievement by matching difficulty to the learner's expertise and gradually increasing the challenge.
- Offer continuous feedback on progress and learning processes.
- Promote self-control by giving learners flexibility in pacing, order, and learning strategies.

Satisfaction: How can learning be rewarding for learners?

- Foster intrinsic reinforcement by encouraging the enjoyment of learning for its own sake and highlighting real-life applications.
- Use extrinsic rewards and positive feedback, preferably at unexpected intervals, in connection with their learning efforts.
- Ensure equity by maintaining high standards and avoiding excessive praise for simple tasks.

Gamification can incorporate many of these motivational factors. Learners are presented with challenges, receive continuous feedback, and are rewarded through points, levels, or competition with peers. This enhances confidence and satisfaction. Rewards such as digital badges or certificates can further boost motivation, making gamification an increasingly considered strategy in digital learning environments.

Digital learning often involves asynchronous self-study, which, due to its flexible nature, places high demands on learners' self-regulation skills. The following tips demonstrate how motivational theories and principles can be translated into practical strategies to enhance student engagement in asynchronous learning.

 **Tips for Engagement in Asynchronous Online Self-study**

- **Clear expectations:** Clearly communicate learning objectives and engagement expectations, including assessment requirements, during self-study.
- **Application:** Use transfer tasks and case studies from learners' professional or personal contexts to engage learners and highlight the relevance of the content.
- **Social learning:** Encourage interaction and collaboration among learners through activities such as collaborative reading assignments using annotation tools or interactions in forums.
- **Status of assignments:** Keep learners updated on assignment progress and highlight notable contributions. Offer completion tracking, send personalized reminders for upcoming or missed deadlines, and provide extensions when appropriate.
- **Test for success:** Provide formative assessments, such as quizzes or tests, to help learners track their progress and experience a sense of accomplishment.
- **Gamification:** Incorporate gamification elements, such as leaderboards, badges, or progress points, to motivate learners as they complete tasks or achieve milestones.
- **Adaptive learning:** Enable personalized learning paths that adjust to learners' individual progress and needs using adaptive learning systems.
- **Integrated learning phases:** Seamlessly integrate online self-study with face-to-face sessions, ensuring that learning activities build upon and reinforce each other.

2.3 Application

As mentioned earlier, a key recommendation of cognitive load theory is to reduce and structure content based on the learner's level of expertise. However, in practice, systematically structured content that follows logical criteria often differs significantly from the complex and less structured demands of everyday professional situations. This can lead to fragmented knowledge (Resnick 1987). As a result, knowledge acquired in the classroom may not be effectively applied in professional contexts, creating a gap between knowledge and action (Renkl et al. 1996). This "transfer problem" has long been an issue in teaching practice; Whitehead (1929) identified and described it as "inert knowledge."

To address this, the learning situation should incorporate key features of the *application* environment, such as including work-related examples and tasks. By doing so, new knowledge stored in long-term memory is linked to real-world applications and can be more easily recalled and applied when needed. The situated nature of knowledge also supports the idea that learning should be centred around authentic, complex problems—those that require cognitive processes similar to solving real-life issues. Examples of such strategies include problem-based learning and the four-component instructional design (4C/ID) model (see Chap. 7), which emphasizes problem-solving and real-world application.

2.4 Design Principles

The following principles for designing digital learning environments are derived from cognitive science:

- *Activate prior knowledge:* Digital learning environments should support the development of comprehensive, well-structured schemata in long-term memory. Activate learners' prior knowledge before introducing new content to help them integrate new information effectively.
- *Focus learner's attention and encourage engagement:* An attractive learning experience that highlights key information, captures learners' interest, and delivers relevant content reduces the likelihood of distraction or dropout.
- *Optimize cognitive load:* Since the human brain has a limited capacity to process information, especially for beginners, avoid unnecessary details. Adjust complex content through segmentation, sequencing, and educational adaptation based on the learner's expertise level.
- *Enhance comprehension through multimodal presentation:* Present information using multiple formats, such as text and images, simultaneously to improve learners' understanding.
- *Promote knowledge transfer through the application context:* To help learners apply knowledge in professional or personal contexts, incorporate realistic scenarios into the digital learning environment. Use problem-solving and decision-making tasks to encourage practical understanding and real-world application.
- *Gradually reduce support:* Use scaffolding techniques, such as worked-out examples or completion tasks, to gradually reduce support and help learners build independence.

You may encounter *conflicting design goals* based on these principles. For instance, reducing the amount and complexity of information aligns with cognitive load limitations. However, providing comprehensive details about authentic application contexts may be essential for promoting motivation and knowledge transfer. This illustrates that no simple recipes can be derived from the concepts and findings of cognitive science. Instead, designing a digital learning environment requires balancing multiple factors, such as the learners' prior knowledge, motivation, learning objectives, and the complexity or emotional nature of the content. Achieving an optimal design involves an iterative process of carefully weighing competing goals based on the specific learning context. The next section presents design models that describe the key steps and factors to be considered.

References

Atkinson, R. C., & Shiffrin, R. M. (1968). Human memory: A proposed system and its control processes. In K. W. Spence & J. T. Spence (Eds.), *Psychology of learning and motivation* (Vol. 2, pp. 89–195). Academic Press. https://doi.org/10.1016/S0079-7421(08)60422-3

Baddeley, A. D. (1986). *Working memory*. Clarendon Press.

Cook, D., & Artino, A. (2016). Motivation to learn: An overview of contemporary theories. *Medical Education, 50*, 997–1014. https://doi.org/10.1111/medu.13074

Eccles, J. S., Adler, T. F., Futterman, R., Goff, S. B., Kaczala, C. M., Meece, J. L., & Midgley, C. (1983). Expectancies, values and academic behaviors. In J. T. Spence (Ed.), *Achievement and achievement motivation* (pp. 75–146). Freeman.

Eccles, J. S., & Wigfield, A. (2020). From expectancy-value theory to situated expectancy-value theory: A developmental, social cognitive, and sociocultural perspective on motivation. *Contemporary Educational Psychology, 61*, 101859. https://doi.org/10.1016/j.cedpsych.2020.101859

Kahu, E. R. (2013). Framing student engagement in higher education. *Studies in Higher Education, 38*(5), 758–773. https://doi.org/10.1080/03075079.2011.598505

Kalyuga, S., Ayres, P. L., Chandler, P., & Sweller, J. (2003). The expertise reversal effect. *Educational Psychologist, 38*, 23–31. https://doi.org/10.1207/S15326985EP3801_4

Keller, J. (1983). Motivational design of instruction. In C. Reigeluth (Ed.), *Instructional design theories and models: An overview of their current studies* (Vol. 1, pp. 383–434). Erlbaum.

Kirschner, P. A. (2017). Stop propagating the learning styles myth. *Computers & Education, 106*, 166–171. https://doi.org/10.1016/j.compedu.2016.12.006

Miller, G. A. (1956). The magical number seven, plus or minus two: Some limits on our capacity for processing information. *Psychological review, 63*, 81–97.

Moreno, R. (2006). Does the modality principle hold for different media? A test of the method-affects-learning hypothesis. *Journal of Computer Assisted Learning, 22*(3), 149–158. https://doi.org/10.1111/j.1365-2729.2006.00170.x

Neelen, M., & Kirschner, P. A. (2020). *Evidence-informed learning design: Creating training to improve performance*. Kogan Page Publishers.

Paivio, A. (1986). *Mental representations: A dual-coding approach*. Oxford University Press.

Park, B., Plass, J. L., & Brünken, R. (2014). Cognitive and affective processes in multimedia learning. *Learning and Instruction, 29*, 125–127. https://doi.org/10.1016/j.learninstruc.2013.05.005

Renkl, A., Mandl, H., & Gruber, H. (1996). Inert knowledge: Analyses and remedies. *Educational psychologist, 31*(2), 115–121. https://doi.org/10.1207/s15326985ep3102_3

Resnick, L. B. (1987). Learning in school and out. *Educational Researcher, 16*(9), 13–20.

Schneider, S., Beege, M., Nebel, S., Schnaubert, L., & Rey, G. D. (2022). The cognitive-affective-social theory of learning in digital environments (CASTLE). *Educational Psychology Review, 34*(1), 1–38. https://doi.org/10.1007/s10648-021-09626-5

Schneider, S., Nebel, S., & Rey, G. D. (2016). Decorative pictures and emotional design in multimedia learning. *Learning and Instruction, 44*, 65–73. https://doi.org/10.1016/j.learninstruc.2016.03.002

Sweller, J., van Merriënboer, J. J. G., & Paas, F. (2019). Cognitive architecture and instructional design: 20 years later. *Educational Psychology Review, 31*(2), 261–292. https://doi.org/10.1007/s10648-019-09465-5

Sweller, J., Van Merrienboer, J. J. G., & Paas, F. G. (1998). Cognitive architecture and instructional design. *Educational Psychology Review, 10*(3), 251–296. https://doi.org/10.1023/A:1022193728205

Vygotsky, L. S. (1978). *Mind in society*. Harvard University Press.

Whitehead, A. N. (1929). *The aims of education*. Williams & Norgate.

Chapter 3
Design Models

Design models and tools provide essential support, particularly for educators new to digital learning. The offer–usage model illustrates the teacher–learner relationship, specifically how the digital learning environment created by the teacher is utilized by learners. Procedural models further support the development of digital learning environments by guiding teachers through key steps that can be enhanced with supplementary tools. The design phase, in particular, is complex due to trade-offs between competing goals. Educational design models help clarify the interaction between core components, deepening the design process and ensuring a more comprehensive learning experience.

This chapter explores the following key questions: What are the key characteristics of teaching and learning situations, and how do they interact? What are the essential steps or phases in developing digital learning environments? Which development approaches have proven successful? How do you design an effective educational experience?

3.1 Offer–Usage Model

Learners organize their learning processes based on their time, available resources, and individual goals and needs, making learning an active and subjective process. While teachers provide a structured learning environment with learning resources, procedural guidance, and social frameworks to direct learners' actions, they cannot dictate how learners engage with these elements. Teachers can design environments and schedules to the best of their knowledge, but the extent to which learners use them is ultimately up to the individual (Goodyear 2015). As Wenger (1998) noted, "First and foremost, learning cannot be designed. Neither can activity nor experience be designed. They can be designed for."

The teacher–learner relationship resembles an *offer–usage* dynamic (Helmke 2003). Teachers provide a learning environment, but learners interpret, adapt, and

© The Author(s) 2025
C. Müller, *Digital Learning Design*, SpringerBriefs in Education,
https://doi.org/10.1007/978-3-031-89045-1_3

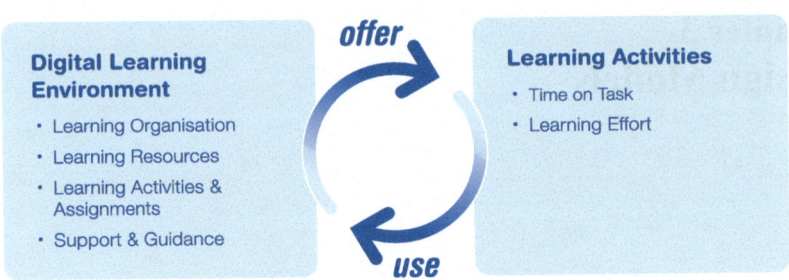

Fig. 3.1 Pedagogical offer–usage model

use it according to their needs (see also Fig. 3.1). Usage depends not only on how much active learning time learners invest but also on how deeply they engage with the learning environment in terms of cognitive, affective, social, and behavioural learning effort (Bond et al. 2020).

This is particularly true for digital learning, where the distance between teacher and learner is greater in blended or online environments. Teachers have fewer opportunities to directly observe or intervene in the learning process, making it harder to guide learners. Teachers must anticipate learners' paths and tailor the learning environment, support, and feedback accordingly. Furthermore, digital learning often combines on-site and online formats, as well as asynchronous and synchronous sessions, adding complexity to the learning experience. Without careful planning, this complexity can overwhelm learners. For instance, a study of award-winning online courses (Martin et al. 2019) found that these courses are characterized by a systematic and thoughtful design.

3.2 Procedural Models and Tools

Procedural models and tools offer essential support, especially when teachers are developing educational designs for unfamiliar contexts, such as blended or online learning. Various instructional systems design models exist, each differing in its level of detail. However, the *ADDIE model* is widely recognized for outlining the key steps in designing and developing digital learning environments (Reigeluth and An 2020). ADDIE includes the phases of *analysis*, *design*, *development*, *implementation*, and *evaluation* (Fig. 3.2).

The *analysis* (phase 1) gathers the necessary information for the subsequent development phases of a digital learning environment. This first phase examines whether there is a need for training at all (needs analysis), who the target group is (target group analysis), and what competencies need to be developed (task and topic analysis). This is to avoid developing a learning environment without taking into account the needs and requirements of the learners. Resources also need to be analysed. For example, some educational designs are suitable for a certain context but cannot always be implemented due to time structures (e.g., the number of lessons at certain intervals) or infrastructure (room characteristics, including ICT equipment).

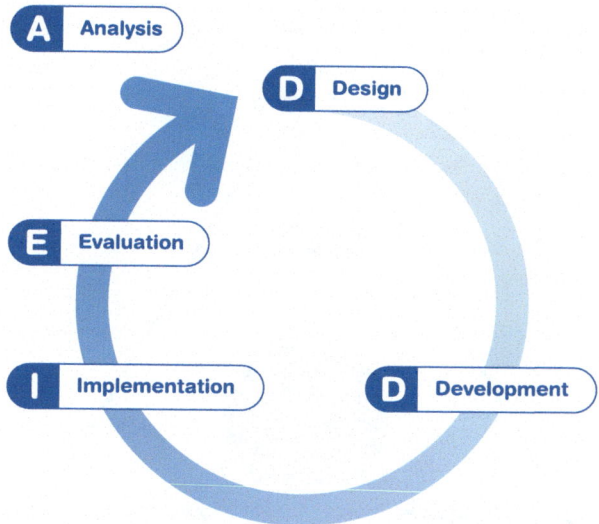

Fig. 3.2 Instructional systems design model, ADDIE

The analysis phase results in the definition of the learning outcomes and, thus, the competencies that the participants should acquire at the end of their engagement in the digital learning environment ("starting with the end in mind"). The pedagogical design is then developed based on the learning outcomes in the sense of *backward design*.

The *design* (phase 2) is dedicated to planning the learning environment. Based on the analysis and the learning outcomes, the learning objectives and content are structured, an appropriate learning organization is determined, and suitable learning strategies are identified. Then, in a circular scripting process, the learning activities are designed with the four aspects of content delivery, activation, interaction, and assessment, and the learning assignments for the activities are planned.

In the *development phase* (phase 3), the required learning resources (e.g., instructional videos and/or texts) are produced, and learning activities are developed and combined into a learning environment. This can be done in an LMS.

During the *implementation phase* (phase 4), the learning environment is launched and supported by guidance, communication, and assistance services. The *evaluation phase (phase 5)* involves a critical review of the learning environment to identify areas for improvement. From an educational perspective, *constructive alignment* (Biggs 1999), the congruence of learning outcomes with the learning environment and assessment, is crucial for evaluation. The learning outcomes serve as guidelines for the development of the learning environment and assessment. In other words, the environment and assessments must support and evaluate the competencies aimed for (see Fig. 3.3).

According to the *"iron triangle"* or "holy grail theory" (Honebein and Reigeluth 2021), other criteria for evaluating learning environments include *learning effectiveness* (outcomes vs. objectives), *learning efficiency* (resources vs. outcomes), and *learning attractiveness* (learner experience and evaluation). However, balancing

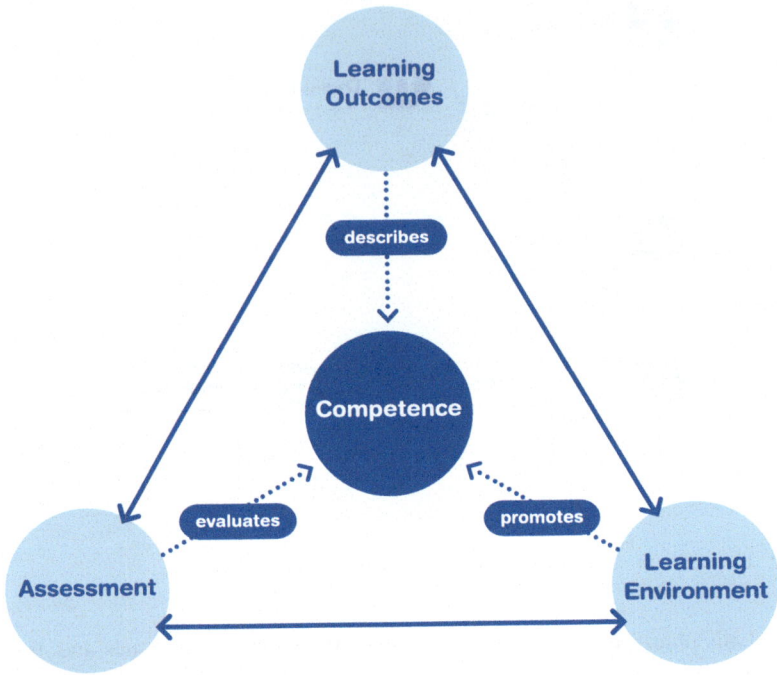

Fig. 3.3 Constructive alignment (Biggs 1999)

these three factors can be challenging, as prioritizing one often impacts another. Focusing on effectiveness and attractiveness may reduce efficiency. For instance, when new tools are introduced, they are often seen as an additional burden by teachers, affecting adoption. Therefore, setting priorities is essential to meeting identified needs as effectively as possible.

The ADDIE model is not intended as a rigid, linear framework but as an *agile process* with feedback loops. For example, after the analysis, it is often helpful to build a prototype for early feedback, refining the design and implementation as needed. Because of the complexity of digital learning environments, optimal implementation is rarely achieved in the first attempt, so recurring evaluations and improvements are crucial.

A flexible, iterative approach also makes sense since creating digital learning environments involves *collaboration among different specialists*, including project managers, SMEs, educational designers, media producers, and learning tool specialists. Iterative feedback loops may promote mutual understanding and improve the overall development process.

A variety of *tools* are available to support the development of digital learning environments. Popular tools for implementation include LMSs such as Moodle, Ilias, Olat, or Canvas; communication tools such as Teams and Zoom; and media platforms such as YouTube. For development, specialized tools, such as authoring platforms (Articulate and Adobe Captivate), video editing tools (Camtasia), and evaluation tools (Limesurvey), are used. The myScripting tool has been specially developed for educational design and is presented in this manual. Other tools, such as Excel or Miro, are also used to design learning environments.

In recent years, AI has become increasingly integrated into teaching and learning tools, with specialized AI applications emerging to support specific tasks in educational design. AI plays a key role in streamlining time-consuming processes, such as context analysis, content creation, the development of interactive and personalized learning activities, and assessment and feedback. By automating repetitive and standardized tasks, AI significantly reduces the workloads of teachers and educational designers. This shift allows educators to concentrate more on planning strategically, designing educational offerings, orchestrating learning activities, and engaging directly with learners—areas where human qualities like intuition, empathy, and creativity are essential. Consequently, the role of educators in digital learning is evolving toward more strategic, creative, and ethical decision-making. While AI can enhance the efficiency and quality of teaching, it complements, rather than replaces, the indispensable role of educators in the learning process.

3.3 Educational Design Models

In the ADDIE process model, the design phase is where the conceptual development of digital learning takes place. Design situations are inherently complex, especially in education, as there are often trade-offs between competing goals and standardized solutions are rarely applicable (Goodyear 2015). For instance, within a single learning unit, multiple objectives may need to be addressed simultaneously, such as teaching specific content, fostering collaboration and communication among learners, and enhancing self-regulation and reflection skills. Optimizing one objective can sometimes hinder the development of another, making it essential to balance these priorities during the design phase.

Furthermore, according to the offer–usage model, it is important to recognize that planned learning activities may not directly translate into actual learning behaviours. Learners interpret, adapt, and enact the activities designed by teachers, often influenced by their individual backgrounds and group dynamics. This highlights the inherent complexity of educational contexts, where no universally "good" or "bad" methods exist and there are no "one size fits all" solutions. Instead, the appropriateness of educational designs and methods depends on the specific context (Honebein and Reigeluth 2021; Tokuhama-Espinosa 2021). Consequently, teachers must thoroughly analyse the context, design appropriate learning environments, anticipate learners' responses, and adapt their support accordingly. This approach aligns with the broader *concept of design* used in other fields (Graham 2019; Laurillard 2013). The design process typically results in sketches, drafts, or blueprints rather than finished products. These educational design artifacts, often called scripts, provide the information necessary to create and implement digital learning environments.

The specific context plays a crucial role in shaping educational design, influencing decisions about learning organization, teaching strategies, and learning activities. Despite this context-specific nature, research on digital learning has identified general principles that can guide the design of effective online and blended learning environments. Four key factors—summarized under the acronym *GAIA*—are particularly critical for digital learning (Müller et al. 2023):

- G (*Guidance*): Adequate course structure and guidance for learners
- A (*Activation*): Activating learning tasks
- I (*Interaction*): Stimulating interactions and the social presence of teachers
- A (*Assessment*): Timely feedback on the learning process and learning outcomes

Among these, *activation* is particularly significant. Teachers often emphasize content delivery (through instructional texts, videos, etc.), but as Merrill (2018) noted, "information alone is not instruction." Learning environments must go beyond content delivery to actively engage learners cognitively, emotionally, and socially. Learning tasks should encourage learners to process information deeply, apply their knowledge in new contexts, and validate their understanding through social interaction and communication. Moreover, studies consistently show that feedback plays a critical role in learning success (Hattie and Timperley 2007). Both formative and summative feedback are essential for guiding learners through their educational journeys.

Figure 3.4 illustrates the interconnectedness of the key components of the *educational design model*. The design of digital learning environments is guided by the results of the analysis and, in particular, by the learning outcomes. Based on these

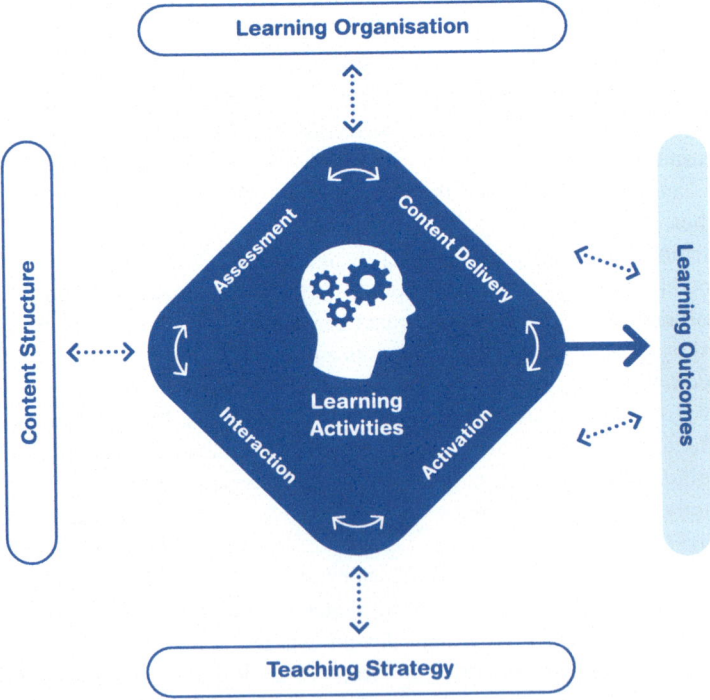

Fig. 3.4 Educational design model

results, the learning organization and content structure are defined, providing a temporal–spatial and content-related framework for the learning experience. Additionally, educational design can be shaped by teaching strategies that are appropriate for the context. At the heart of the educational design model are the learning activities, which are divided into four groups: content delivery, activation, interaction, and assessment.

The educational design model demonstrates that all design aspects are closely interrelated; therefore, the development process is not linear. Continuous evaluation of the digital learning environment in terms of constructive alignment and the iron triangle (effectiveness, efficiency, and attractiveness), along with ongoing optimization of the educational design, is crucial. Chapters 4 (Analysis), 5 (Learning Organization), 6 (Content Structure), 7 (Teaching Strategies), and 8 (Learning Activities) provide more detailed descriptions of the individual aspects of the educational design model.

 **Tips for Educational Design Process**

- **Backward design:** Begin by determining the desired learning outcomes—what competencies learners should achieve by the end of the course ("beginning with the end in mind"). Use these outcomes to guide content structuring and develop educational design.

- **Constructive alignment:** Ensure that both the designed learning environment and the assessment strategies are aligned with the learning outcomes.

- **GAIA principle:** Apply the GAIA framework as guidelines for designing effective digital learning environments. Provide clear instructions, learning paths, resources, and support (guidance); develop tasks that actively engage learners in thinking and application (activation); encourage interaction and collaboration through discussions, group work, or virtual meetings (interaction); and offer timely feedback to help learners monitor their progress and improve (assessment).

- **Balanced design (Iron Triangle):** Strive for a balance between learning effectiveness, efficiency, and attractiveness. A highly effective but unengaging learning experience can reduce motivation, while an overly attractive but inefficient course may waste time. Prioritize your design goals based on learner needs and available resources.

- **Agile design process:** Adopt an agile, iterative approach to developing digital learning environments, incorporating regular feedback loops to refine and improve the design continuously.

myScripting   **Educational Design in myScripting**

The design process with myScripting is based on the educational design model described above (see Figure 3.4). In the first step, the temporal-spatial learning organization (horizontal) and the content structure (vertical) are developed based on the analysis. Subsequently, the learning activities—content delivery, activation, interaction, and assessment—are aligned in a circular educational design process. These activities can be guided by specific teaching strategies, such as direct instruction or problem-based learning.

Although myScripting focuses on the design phase of digital learning environments, the tool also supports the preceding (analysis) and subsequent phases (development, implementation, evaluation) of the ADDIE model:

- *Analysis:* The results of the analysis are documented in myScripting under "Prerequisites" (e.g., regarding learners), "Learning Outcomes," "Content," and "Assessment." The analysis also guides the basic script settings, such as "Workload," "Target platform," and "Form of assessment."

- *Development and Implementation:* The developed script can be exported to an LMS for the creation of a digital learning environment. Action-guiding and role-specific outputs for the teaching and learning processes can also be generated from the scripts. For teachers, a chronological lesson plan is created in a table view containing the most important information for conducting the lesson; learners receive a syllabus (course overview and summary).

- *Evaluation:* A review can be conducted directly in myScripting, with feedback provided for individual learning phases, topics, and the entire script. The respective development status can also be tracked.

References

Biggs, J. B. (1999). *Teaching for Quality Learning in University.* Society for Research in Higher Education and Open University Press.

Bond, M., Buntins, K., Bedenlier, S., Zawacki-Richter, O., & Kerres, M. (2020). Mapping research in student engagement and educational technology in higher education: A systematic evidence map. *International Journal of Educational Technology in Higher Education, 17*(1), 2. https://doi.org/10.1186/s41239-019-0176-8

Goodyear, P. (2015). Teaching as design. *Herdsa review of higher education, 2*(2), 27–50.

Graham, C. R. (2019). Current research in blended learning. In M. G. Moore & W. C. Diehl (Eds.), *Handbook of distance education* (4th ed., pp. 173–188). Routledge.

Hattie, J., & Timperley, H. (2007). The power of feedback. *Review of Educational Research, 77*(1), 81–112. https://doi.org/10.3102/003465430298487

Helmke, A. (2003). *Unterrichtsqualität - erfassen, bewerten, verbessern*. Kallmeyer.

Honebein, P. C., & Reigeluth, C. M. (2021). Making good design judgments via the instructional theory framework. In M. J.K. & R. E. West (Eds.), *Design for Learning: Principles, Processes, and Praxis*. https://open.byu.edu/id/making_good_design

Laurillard, D. (2013). *Teaching as a design science: Building pedagogical patterns for learning and technology*. Routledge.

Martin, F., Ritzhaupt, A., Kumar, S., & Budhrani, K. (2019). Award-winning faculty online teaching practices: Course design, assessment and evaluation, and facilitation. *The Internet and Higher Education, 42*, 34–43. https://doi.org/10.1016/j.iheduc.2019.04.001

Merrill, M. D. (2018). Using the first principles of instruction to make instruction effective, efficient, and engaging. In R. E. West (Ed.), *Foundations of learning and instructional design technology: The past, present, and future of learning and instructional design technology*. EdTech Books. https://edtechbooks.org/lidtfoundations/using_the_first_principles_of_instruction

Müller, C., Mildenberger, T., & Steingruber, D. (2023). Learning effectiveness of a flexible learning study programme in a blended learning design: Why are some courses more effective than others? *International Journal of Educational Technology in Higher Education, 20*(1), 10. https://doi.org/10.1186/s41239-022-00379-x

Reigeluth, C. M., & An, Y. (2020). *Merging the instructional design process with learner-centered theory: The holistic 4D model*. Routledge. https://doi.org/10.4324/9781351117548

Tokuhama-Espinosa, T. (2021). *Bringing the neuroscience of learning to online teaching: An educator's handbook*. Teachers College Press.

Wenger, E. (1998). *Communities of practice: Learning, meaning, and identity*. Cambridge University Press. https://doi.org/10.1017/CBO9780511803932

Chapter 4
Context Analysis

The educational setting is carefully examined to gather all the essential information required for designing an effective digital learning environment (see Fig. 4.1). Identifying training needs ensures alignment with learners' characteristics and requirements, preventing designs that might overlook critical needs. Resource assessment is equally crucial; even if a learning solution is well suited to the context, factors such as scheduling (e.g., lesson frequency and duration) or infrastructure (e.g., room setup and ICT resources) could present challenges to implementation. Based on insights from the context analysis, the intended learning outcomes are then defined.

This chapter addresses key questions: Is there a need for training? Who is the target audience, and what are their primary characteristics? Which competencies should be developed, and what learning outcomes are targeted? What resources are available to support the creation of a learning environment?

Fig. 4.1 Analysing the educational context

© The Author(s) 2025
C. Müller, *Digital Learning Design*, SpringerBriefs in Education,
https://doi.org/10.1007/978-3-031-89045-1_4

4.1 Needs Analysis

Before designing a digital learning environment, a *needs analysis* is carried out to determine whether training is needed at all and whether a digital learning environment is the best solution. Particularly in companies, a performance problem can have various causes, and they may not necessarily relate to insufficient knowledge and skills but rather to inadequacies in the working environment, such as insufficient information, work equipment, or incentives (Mager and Pipe 1984).

Additionally, not all learning needs are equally suited to digital learning. Digital learning works best under specific conditions, such as when (e.g., McKenna et al. 2020):

- There is a significant amount of content to be learned.
- There are a large number of learners.
- Learners are geographically dispersed.
- Learners face mobility constraints.
- Learners have limited time per day to devote to learning.
- Learners possess basic digital literacy and internet access.
- Learners appreciate learning at their own pace.
- The course content can be reused across multiple learner groups.
- The course focuses on the development of professional competence.
- Proof of training and learning outcomes is needed (e.g., for compliance).

However, there are also situations where e-learning may not be the best fit, particularly when:

- The course relies heavily on learner interaction and collaboration.
- Learners' diverse knowledge and experience levels require adaptive, responsive teaching.
- The course involves practical skills that require special equipment or authentic settings.
- Networking and social interaction are key course benefits.

If there is a need for training or education, and if it is also suitable for digital learning, the next steps are a more detailed analysis of the target group and the definition of the learning outcomes.

4.2 Target Audience Analysis

Understanding the *target audience* is essential for designing an effective digital learning environment. Key characteristics to consider include learners' prior competency levels, access to technology and ICT literacy, and their motivation to engage with the content. Factors such as age, gender, and educational background also play a role in shaping the design.

Important questions when analysing the target audience include the following:

- How many learners will participate?
- What relevant competencies do learners already possess?
- How homogeneous or heterogeneous is the group in terms of competency levels?
- What prior experiences can be leveraged in the course?
- What is their motivation for the content?
- What general learning skills do learners have?
- What experience do they have with digital learning environments?
- What access do they have to ICT?
- How flexible are learners in terms of time and location?

4.3 Task and Topic Analysis

When designing a digital learning environment, it is essential to define the competencies that learners need to develop. If an existing course is adapted to digital learning, the learning outcomes can often be derived from the course syllabus or objectives. However, when developing a new course, a detailed *competency analysis* is essential to determine these outcomes. Identifying the required competencies involves analysing the skills, knowledge, and attitudes necessary for success in the professional or occupational field (see Table 4.1). This process often includes

Table 4.1 Competence grid (based on Baumgartner et al. 2018)

Competency	Description and subcompetencies
Professional competency	Professional competencies are the specific abilities needed to master subject contents of theoretical and practical relevance.
Methodological competency	Methodological competencies are the abilities that can be applied across different situations and are needed to meet difficult challenges in the workplace. These include: • Problem-solving • Abstract and networked thinking • Digital literacy • Foreign language skills
Social competency	Social competencies are abilities that are needed to be effective in reaching professional goals in situations of social interaction. These include: • Ability to cooperate • Oral and written communication skills • Teamwork and conflict management skills
Self-competency	Self-competencies are abilities and attitudes that are needed to develop professionally and actively engage and be effective in a workplace environment. These include: • Self-management (e.g., stress management and motivation to learn) • Ethical and social responsibility • Identity development (e.g., self-concept and the ability to self-critique)

observations and interviews to assess the capabilities needed to perform specific tasks or handle workplace situations (Baumgartner et al. 2018). When analysing tasks, a distinction can be made between procedural and principle-based tasks:

- *Procedural tasks* require actions to be performed in a specific order. Examples include creating a formula in Excel or assembling a piece of furniture. The steps required for the procedure can be identified in the analysis, and the required sub-competencies can be determined.
- *Principle-based tasks* require judgements and decisions in changing contexts. An example of this is managing or facilitating groups. Precise procedures cannot be defined, but certain actions that are useful in a particular context can be identified and summarized in guidelines. The necessary subcompetencies can then be identified.

For courses that focus on delivering information, a *topic analysis* is needed. This helps in organizing and structuring the course content, ensuring it is logically sequenced (see Chap. 6 for more on content structuring).

4.4   Resource Analysis

Resource analysis helps clarify the infrastructure, human resources, and content resources available to design the learning environment. To guide this process, consider the following key areas and questions:

Human resources—Teachers

- What time is allocated for the development and implementation of the learning environment?
- What expertise (professional, educational, or technical) is available for its development?

Time resources—Learners

- How many hours of learning time (workload) are planned for the learning outcomes?
- What time structures are available (e.g., hours per week and block schedules)?
- What is the expected duration for learners to complete the course?

Teaching facilities

- What types of classrooms are available (size, layout, and flexibility of equipment)?
- Are there specialized facilities (e.g., laboratories and robotics rooms) to support skill development?
- How accessible are the teaching buildings (location, public transport, and parking)?
- How appealing is the campus environment (e.g., cafeteria, lounge areas, and other services)?

Technical infrastructure

- How are the classrooms equipped (e.g., projector, smart board, and hybrid learning facilities)?
- What learning tools are available (e.g., LMS and collaborations tools)?
- What technical equipment do learners have (e.g., laptops, tablets, and media design tools)?

Content resources

- What learning materials are already available or can be procured in a timely and legally compliant manner (e.g., teaching texts, media)?

In terms of *financial considerations*, developing a digital learning environment, particularly an asynchronous one, often requires a significant initial investment. However, asynchronous environments can offer substantial cost benefits in the long run, as they are particularly efficient due to their reproducibility and scalability—especially when intended for large or multiple learner groups.

4.5   Learning Outcomes

Learning outcomes are determined based on the results of the analysis of the target group and the analysis of tasks, content, and resources. Learning outcomes describe the competencies that learners should possess at the end of the learning process. Learning outcomes make a statement that is as observable as possible about the type and differentiation of the desired behaviour in a specific context.

Therefore, the level of complexity of the learning outcomes depends on two factors: the complexity of the content and the nature and differentiation of the required behaviour. Whether content is complex must be assessed from the perspective of the discipline. It depends on the number of information units to be processed and their complexity. In terms of behaviour, the *cognitive taxonomy of Krathwohl and Anderson* (2009) is often used.

Learning outcomes should specify externally observable and measurable behaviours of learners, making them operational and practical to assess. For instance, using action verbs such as "name" and "describe" is preferable to terms such as "know" and "understand," as the latter are not directly observable. Additional recommended verbs for formulating effective learning outcomes, aligned with Krathwohl and Anderson's (2009) taxonomy, are listed in Table 4.2 (based on Newton et al. 2020).

It is important to note that the level of cognitive demand also depends on whether the behaviour has been practiced during prior learning sessions. For example, if a task has been discussed in class, learners may memorize the answer. If a similar task is given during an exam, it becomes a simple recall task at the "remember" level. Thus, starting from the "apply" level, learners must demonstrate the behaviour in a new context.

Table 4.2 Formulation of learning outcomes

Remember	List, define, recall, state, label, repeat, name
Understand	Translate, paraphrase, discuss, report, locate, generalize, explain, classify, summarize
Apply	Operate, apply, use, demonstrate, solve, produce, prepare, choose
Analyse	Analyse, question, differentiate, experiment, examine, test, categorize, distinguish, calculate, contrast, outline, infer, discriminate, compare
Evaluate	Rate, evaluate, assess, judge, justify
Create	Create, compose, argue, design, plan, support, revise, formulate

When formulating effective learning outcomes, the *SMART* criteria provide a structured approach that enhances clarity and focus. This ensures learners and educators have a shared understanding of the intended behaviour.

- *Specific:* Outcomes must be clearly defined and concrete, using precise language to avoid vague or abstract terms.
- *Measurable:* Each outcome should include observable actions that can be assessed through learner behaviour, specifying what learners will demonstrate or perform.
- *Achievable:* Outcomes should be realistically attainable within the given constraints, such as available time, resources, and the learner's current competency level.
- *Relevant:* Outcomes should be meaningful, engaging, and challenging enough to sustain motivation while still being manageable for the learner.
- *Time-bound:* Outcomes should include a timeframe for achievement, indicating when learners are expected to meet goals.

In some contexts, certain SMART elements, such as a specific timeframe, may already be established (e.g., course or semester duration). In these cases, it may not be necessary to repeat such details in the outcome formulation, as they are implicitly understood.

 Tips for Context Analysis

- **Learning outcomes:** Define and operationalize the learning outcomes in a differentiated way. These are crucial for the educational design but also for the evaluation of the digital learning environment.
- **Target group analysis:** Analyse the learners' requirements and needs in detail. These are guiding factors for the temporal–spatial learning organization of the digital learning environment.
- **Conflicting goals:** On the basis of the context analysis and the available resources in particular, weigh the goals of the digital learning environment in terms of learning effectiveness, learning efficiency, and attractiveness of the learning environment, and then set meaningful priorities.

myScripting **Context Analysis in myScripting**

The results of the context analysis are documented in myScripting in the fields "Preconditions" (e.g., regarding learners), "Learning outcomes," "Content," and "Assessment." Further sources and features of the learning environment can be documented in the extended information of the script. Context analysis is also used to determine the following three general script settings: planned workload, target platform, and assessment system.

The topics and learning activities can be assigned to the learning outcomes in myScripting. The tool then analyses the number and workload of topics, subtopics, and learning activities that promote different learning outcomes.

References

Baumgartner, A., Müller, C., Fengler, R., & Javet, F. (2018). Development of application-oriented competency frameworks: Empirical findings from the validation of such a framework by means of an employer survey. *The Journal of Competency-Based Education, 3*(4), e01177. https://doi.org/10.1002/cbe2.1177

Krathwohl, D. R., & Anderson, L. W. (2009). *A taxonomy for learning, teaching, and assessing: A revision of Bloom's taxonomy of educational objectives.* Longman.

Mager, R. F., & Pipe, P. (1984). *Analyzing performance problems* (2. ed.). Lake Publishing.

McKenna, K., Gupta, K., Kaiser, L., Lopes, T., & Zarestky, J. (2020). Blended learning: Balancing the best of both worlds for adult learners. *Adult Learning, 31*(4), 139–149. https://doi.org/10.1177/1045159519891997

Newton, P. M., Da Silva, A., & Peters, L. G. (2020). A pragmatic master list of action verbs for bloom's taxonomy [Brief Research Report]. *Frontiers in Education, 5.* https://doi.org/10.3389/feduc.2020.00107

Chapter 5
Learning Organization

The rapid development of digital media and the means of communication in recent years have created a multitude of possibilities for designing learning opportunities in terms of time and space. When deciding on a learning organization, the needs of the learners should be the primary consideration. However, these can vary and may require different formats or hybrid models tailored to individual learning preferences. In addition to learner needs, other factors must also be considered, such as the desired learning outcomes and the existing infrastructure. Therefore, certain learning contexts may require an on-site physical presence for pedagogical or logistical reasons. For example, for skills training that requires specialized equipment or laboratories, the presence of learners is essential.

Last but not least, the way learning is organized also depends on the strategic orientation of the educational institution. The desired learning culture can be a guiding principle for the entire temporal and spatial design of a learning offering and may shape the planned learner journey, as well as the desired learning experiences.

This chapter explores the following key questions: What time and space formats are available? What are the pros and cons of asynchronous versus synchronous learning? What are the features of blended and online learning?

5.1 Time and Space

Based on the analysis and the learning outcomes developed, the dimensions of *time* and *space* are determined for the learning organization. In terms of space, a distinction is made between *on-site* (e.g., on campus) and *online* (virtual—now mainly on the internet). In terms of time, *synchronous* means that learners' learning activities take place live, that is, with the learners simultaneously. In *asynchronous* learning,

C. Müller, *Digital Learning Design*, SpringerBriefs in Education,
https://doi.org/10.1007/978-3-031-89045-1_5

learning activities take place at different times and at the learner's own pace, allowing for more learning flexibility.

A *physical presence event* refers to face-to-face instruction, usually in the
classroom. Other learning locations are also possible, such as learning in the
field, in a laboratory, or in a professional environment, such as a workshop or
hospital. Like the learning centre or laboratory, these alternative learning locations are also used for *on-site self-study*. For a long time, e-learning was primarily implemented in the form of an asynchronous *online self-study*. With the
development of online communication and collaboration tools, the technical
and pedagogical possibilities for *virtual presence events* have expanded. Digital
tools have enhanced the scope of virtual presence events by adding features
such as breakout rooms, which facilitate various social learning formats (e.g.,
plenary, group, or pair activities).

Figure 5.1 shows an overview of the possibilities offered by the dimensions of
time and space.

When choosing a temporal structure, it is essential to consider the target audience, the skills to be acquired, and the technological infrastructure available.
Table 5.1 summarizes the main advantages and disadvantages of synchronous and
asynchronous learning. It should be noted that asynchronous learning today is usually implemented in a digital learning environment. In the past, this was also the
case with physical learning materials (e.g., textbooks and teaching materials for
self-study).

Modern infrastructure also allows for hybrid learning, in which on-site and
online learners participate simultaneously. These sessions are often recorded for
later asynchronous use, a model known as *hyflex* courses. The main features of the
synchronous learning formats are listed in Table 5.2.

Time / Space	Synchronous	Asynchronous
On-site	**Physical presence event** (e.g., in classrooms, in the field, or in the laboratory)	**On-site self-study** (e.g., in the learning centre or in the laboratory)
Online	**Virtual presence event** (e.g., with video conferencing systems)	**Online self-study** (using electronic learning resources)

Fig. 5.1 Learning formats depending on time and space

Table 5.1 Asynchronous versus synchronous learning

Asynchronous learning	Synchronous learning
✓ Allows flexible learning according to one's own schedule and compatibility of learning with other commitments (work, family, and leisure) ✓ The learning process (speed and rhythm) can be adapted according to prior knowledge/language skills/needs or the level of difficulty of the learning content ✓ Personal skills (self-management and self-reflection) are necessary for self-regulated learning and are promoted	✓ Enables direct interaction and feedback between learners and teachers ✓ Gives teachers a better overview and closer guidance of the learning process ✓ Promotes social inclusion and group identity through more authentic social interactions (including outside the classroom)
✗ Learning can be postponed (procrastination), and learners can fall behind as a result ✗ There may be a sense of social isolation due to less direct social interaction	✗ Simultaneous participation may be difficult due to different schedules or time zones ✗ Learners with disabilities (language, disability, prior knowledge, or cognition) may not reach their full potential, as learning is more instructor-paced

Table 5.2 Forms of synchronous learning

Synchronous on-site	Synchronous online	Hybrid
✓ Use of special physical learning equipment (e.g., laboratory, VR equipment, studio, or robot) ✓ Facilitates interaction and networking ✓ Investment/commitment of learners to travel to campus can lead to higher learning engagement/lower dropout	✓ Allows spatial flexibility and therefore better compatibility with other commitments ✓ Shorter attendance units are possible (e.g., one-hour webinars) ✗ Distractions and low exit costs can encourage dropout	✓ The form of presence can be chosen according to the needs of the learners ✗ Designing interactions is challenging for teachers

5.2 Blended Learning

Blended learning integrates synchronous sessions with asynchronous self-study, requiring careful coordination of content and timing between phases. Since synchronous time is often limited, it should be used strategically. Rather than delivering lectures, in-person sessions are best suited for clarifying complex topics, facilitating discussions, promoting interaction, and engaging in group activities that apply learning (Martin et al. 2023). These sessions can also introduce the learning process, familiarize learners with the technology, and build group cohesion, such as during a kick-off for group work. In contrast, the asynchronous phase is ideal for delivering content, individual practice, progress assessments, and reflective activities, such as forum discussions or peer feedback. Common blended learning formats are outlined in Table 5.3 (see also McKenna et al. 2020).

Table 5.3 Models of blended learning

Models	Description
Replacement model	Replaces part of the synchronous classroom time with asynchronous online learning
Flipped classroom	Content is absorbed through self-study, often using explanatory videos, while in-person time is reserved for clarifying questions, applying knowledge, interaction, and assessment
Preparation model	Ensures learners have a consistent knowledge level through preparatory online learning
Training model	Supplements classroom instruction with online learning to practice and apply the content taught
Skills model	Combines online learning with classroom-based skills training
Blended MOOC	Uses freely available massive open online courses (MOOCs) complemented by on-site or online presence sessions

The *flipped classroom* model has gained popularity in educational institutions in recent years, but learning outcomes vary widely (e.g., Kapur et al. 2022). This may be due to the way the model is implemented, as teachers often report that learners' self-study preparation is inadequate or superficial (e.g., Müller et al. 2023). To address this, principles for enhancing learner motivation should be carefully considered when designing the self-study component of the flipped classroom (see also tips for engagement in asynchronous online self-study in Sect. 2.2).

5.3   Online Learning

In online learning, asynchronous self-study is typically emphasized. An important consideration when designing an online course is whether it will be *self-paced* or *instructor-paced*. In a self-paced course, learners control their own progress and complete the material at their own speed. In an instructor-paced course, the instructor releases content incrementally (usually weekly or biweekly). The key characteristics of each course type are summarized in Table 5.4.

When online courses are freely accessible, they are commonly referred to as *massive open online courses (MOOCs)*. MOOCs are often initially launched as instructor-paced courses to allow adjustments based on learner feedback during the first offering, before being made available as self-paced courses.

Table 5.4 Online course: instructor-paced versus self-paced

Instructor-paced	Self-paced
– Course starts at fixed times	– Course start can be chosen
– The pace of learning is set by instructor	– The pace of learning is individual
– Interaction between learners is possible (e.g., forums, webinars, and peer assessment)	– Interaction between learners is limited
– Learners are supported and assessed in cohorts	– Learner support and assessment is individual but time consuming
– Content and activities can be adjusted during the course (based on learners' progress or needs)	– Content and activities are fixed once the course begins (the entire course is available upfront)

 Tips for Learning Organization

- **Prioritize learner needs:** Focus on the requirements of learners, particularly in adult education, when designing the temporal and spatial organization of the learning environment.

- **Balance learning modes:** Combine synchronous and asynchronous learning with on-site and online formats based on learner preferences and course objectives.

- **Integrate self-study and face-to-face learning seamlessly:** Design the learning experience so that asynchronous online learning and synchronous learning events complement each other with activities that build on each other.

- **Maximize face-to-face time:** Use face-to-face time—whether online or on-site—for clarifying, applying, and discussing complex concepts. Focus on fostering interaction and collaboration to build a strong learning community.

- **Promote engagement:** Offer incentives and support to boost engagement in both asynchronous and synchronous online learning activities.

myScripting — Learning Organization in myScripting

In myScripting, based on certain influencing factors, the following decisions are made about the structural features of the learning organization:

Structural features	Influencing factors
Proportion of asynchronous/synchronous learning	Learner availability and needs, and desired learning culture
Proportion of online vs. on-site learning	Learner availability and needs, required and available learning infrastructure (e.g., laboratory facilities), and desired learning culture
Workload	Planned (formal education) or reasonable workload (informal education), and content structure
Length of learning units	Learner availability and needs, and desired learning culture
Control (self-paced or instructor-paced)	Learner availability and needs, instructors' resources, and desired learning culture

Based on the above decisions, an initial structure of the planned learning phases is defined. For synchronous learning, myScripting provides the learning phases' physical presence event and virtual presence event, and for asynchronous learning, it provides the learning phase self-study. The learning phases can be adapted during the design process.

References

Kapur, M., Hattie, J., Grossman, I., & Sinha, T. (2022). Fail, flip, fix, and feed – Rethinking flipped learning: A review of meta-analyses and a subsequent meta-analysis [Systematic Review]. *Frontiers in Education, 7.* https://doi.org/10.3389/feduc.2022.956416

Martin, F., Kumar, S., Ritzhaupt, A. D., & Polly, D. (2023). Bichronous online learning: Award-winning online instructor practices of blending asynchronous and synchronous online modalities. *The Internet and Higher Education, 56,* 100879. https://doi.org/10.1016/j.iheduc.2022.100879

...

McKenna, K., Gupta, K., Kaiser, L., Lopes, T., & Zarestky, J. (2020). Blended learning: Balancing the best of both worlds for adult learners. *Adult Learning, 31*(4), 139–149. https://doi.org/10.1177/1045159519891997

Müller, C., Mildenberger, T., & Steingruber, D. (2023). Learning effectiveness of a flexible learning study programme in a blended learning design: Why are some courses more effective than others? *International Journal of Educational Technology in Higher Education, 20*(1), 10. https://doi.org/10.1186/s41239-022-00379-x

Chapter 6
Content Structuring

In a learning offering, the content is determined through task and topic analysis, while the level of abstraction (overview or detailed) and the desired complexity are determined by defining the learning outcomes. As discussed in the Principles of Cognitive Science chapter, the limited capacity of working memory makes it difficult for learners to fully absorb content and its interconnections at once. To address this, the content must be prioritized, reduced, or simplified *(educational adaptation)*, segmented into manageable parts *(segmentation)*, and presented in logical order *(sequencing)* (see Fig. 6.1).

This chapter explores the following key questions: How can learning content be adapted educationally? What principles can be used for segmenting and sequencing content?

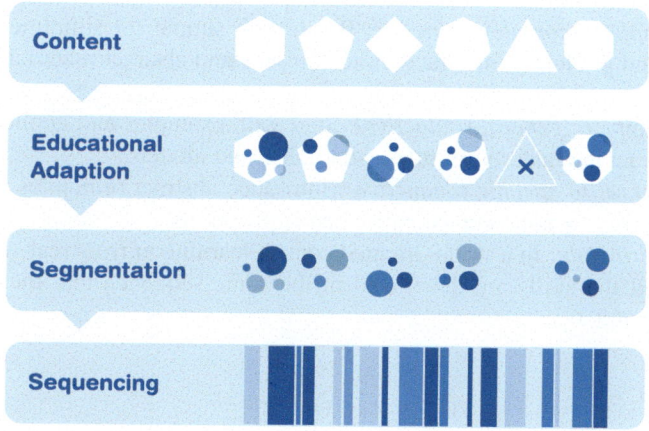

Fig. 6.1 Content structuring process

6.1 Educational Adaptation and Knowledge Structure

Educational adaptation focuses on prioritizing and reducing content to make it manageable and understandable for learners. Prioritization involves selecting key content and omitting nonessential material (content reduction). Additionally, qualitative reduction simplifies content (difficulty reduction) while ensuring that it remains meaningful and applicable. Care should be taken to avoid trivializing the material so that it remains valid for the intended application context.

In practice, *knowledge structures* have been developed to reduce information to its essential elements, aligning with educational adaptation to minimize cognitive load (see Table 6.1).

6.2 Content Segmentation and Sequencing

The aim of *content segmentation and sequencing* is to structure content in an educationally meaningful way. At this stage, it is not about designing the learning environment through learning activities such as exercises, interactive forums, or tests—that comes later in the educational design process. A widely used approach to segmenting and sequencing content is the *learning hierarchy*, in which foundational content is first introduced to ensure that learners have the necessary knowledge or skills to tackle more advanced topics. This sequential approach builds a scaffolded learning experience, allowing each new concept or skill to build on the previous one. However, there are several other *principles for content segmentation and sequencing* content (see also Reigeluth 1999):

- *From simple to complex principle*: Begin with simple (or simplified) concrete content and gradually introduce more complex and abstract material as learners progress.
- *From specific to general (inductive)*: Start with examples and applications, and then abstract the underlying rules, principles, and models.
- *From general to specific (deductive)*: Introduce abstract principles and models first, followed by concrete examples and applications.
- *Process principle:* In a skills-oriented course, learning mirrors real-world procedures, and the skills are developed in the same sequence that they would be applied in practice.

Table 6.1 Knowledge structures for representing information

Type of problem	Challenge	Example
Description	What is something made of? How is something put together?	List, diagram, or disposition
Classification	What/who is superordinate? What/who is subordinate? What are the main ideas? What are the supporting items for each main idea?	Tree, delta, and concept map
Comparison	What are the similarities/differences? What criteria can be used to judge something? Which elements (objects, solutions, alternatives, etc.) are to be compared or assessed?	Matrix
Relationships	How do different things relate to each other? What is dependent on what?	Network and graphics
Process/Sequence	What happens step by step? In what order do you do something? What is the result?	Timeline, flowchart, and schedule

- *Chronological principle:* Content is structured in the order of historical or chronological events.
- *From familiar to new principle:* If learner profiles (e.g., characteristics, professional experience, and educational background) are known, begin with familiar content before introducing more unfamiliar or complex material.
- *Zoom principle*: The course starts with a broad overview and then focuses on specific topics before concluding with a general summary.

In designing a comprehensive learning environment, for example, the curriculum of a study program, the sequence in which topics are presented plays a crucial role in facilitating effective understanding and retention. Two widely recognized approaches to sequencing are the *topical* and *spiral curricula* (Reigeluth 1999), each offering distinct advantages and challenges:

- *Topical curriculum*: In a topical curriculum, each topic is covered thoroughly before moving on to the next. This approach allows learners to focus deeply on each concept, fostering a strong grasp of the material within each segment. However, this can lead to challenges in recalling earlier content or connecting ideas across topics, as there is little reinforcement of previous material. Learners may struggle to see the broader connections between topics, which can hinder comprehensive understanding.

- *Spiral curriculum:* In a spiral curriculum, topics are revisited and explored in increasing depth across multiple stages, allowing learners to gradually build a complete understanding of individual topics and their interconnections. This spaced sequencing encourages long-term retention, as the content is reinforced over time. However, this approach requires learners to adapt to a more interrupted style of progression, which can disrupt their comprehension flow.

Additionally, learners unfamiliar with the spiral curriculum may see repetition as redundant, potentially impacting motivation.

Spiral Sequencing

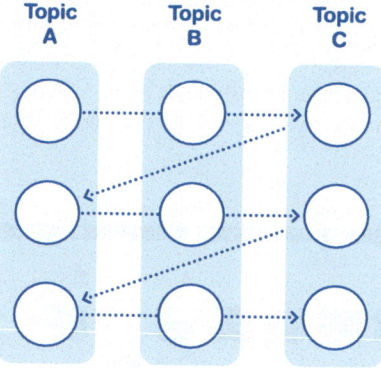

Interleaving also involves the repetition of content, although here, the topics are not learned in sequence but rather in related but not identical contexts, interleaved with each other (Neelen and Kirschner 2020). This method enhances comprehension by allowing learners to encounter topics in varied contexts, reinforcing understanding through repeated exposure (Yan and Sana 2021). However, organizing and sequencing content within an interleaved curriculum can be challenging. Inquiry learning approaches, such as problem-based or project-based learning, naturally incorporate cross-curricular elements and can help coordinate interleaving in the curriculum.

For highly complex content with many interacting elements, cognitive load can be reduced by breaking the learning process into substeps and isolating individual elements. Once these subelements are understood, they are integrated to form a complete understanding (e.g., 4C/ID model in Chap. 7, "Teaching Strategies").

 **Tips for Content Structuring**

- **Simplify without trivializing:** Prioritize key concepts, break content into manageable chunks, and reduce complexity while preserving its practical value.
- **Leverage existing knowledge structures:** Organize content using frameworks and structures that are already well established in the subject area.
- **Interleaving:** Reinforce learning by repeating content in varied contexts, gradually increasing its complexity.

myScripting   **Content Structuring in myScripting**

In myScripting, the content structure for each topic is organized vertically (see the designer view below). Each row represents a topic, mapping out the learning process and associated activities. Subtopics are managed by grouping the related learning activities within each row. Various sequencing patterns, such as epochal or spiral curricula, can be created within myScripting.

References

Neelen, M., & Kirschner, P. A. (2020). *Evidence-informed learning design: Creating training to improve performance*. Kogan Page Publishers.
Reigeluth, C. M. (1999). The elaboration theory: Guidance for scope and sequence decisions. In C. M. Reigeluth (Ed.), *Instructional-design theories and models: A new paradigm of instructional theory* (Vol. 2, pp. 425–453). Lawrence Erlbaum Associates Publisher.
Yan, V. X., & Sana, F. (2021). The robustness of the interleaving benefit. *Journal of Applied Research in Memory and Cognition*, 10(4), 589–602. https://doi.org/10.1016/j.jarmac.2021.05.002

Chapter 7
Teaching Strategies

Teaching strategies provide structured sequences and designs for learning activities, serving as essential models for creating effective digital learning experiences. A clear understanding of the characteristics of these strategies enables the selection of approaches best suited to specific learning contexts and objectives. This chapter explores various teaching strategies, offering insights into how different approaches can enhance student engagement, support diverse learning needs, and promote meaningful outcomes.

This chapter explores the following key questions: What are the main characteristics of the different teaching strategies? Which strategies are most appropriate for specific educational contexts?

7.1 Direct Instruction Versus Inquiry Learning

There are many teaching strategies, and they are generally grouped into two educational models: direct instruction and inquiry learning. These models differ in the design of the learning environment, learning process, and roles of teacher and learner.

In *direct instruction*, the teacher actively guides the learning process, providing explanations and instructions and arranging phases for content delivery, practice and application, and assessment. In direct instruction, learning tends to be linear and systematic, corresponding to a highly externally controlled receptive process. This model is often referred to as explicit teaching, in which the focus is on instructional issues and the teacher is mainly active in presenting, guiding, explaining, and evaluating.

Inquiry learning summarizes approaches to learning in which new content needs to be explored, understood, and applied on the basis of problem situations. Inquiry learning is a more multidimensional, less systematic, and primarily self-directed constructive process. Inquiry learning, also known as implicit learning or discovery

© The Author(s) 2025
C. Müller, *Digital Learning Design*, SpringerBriefs in Education,
https://doi.org/10.1007/978-3-031-89045-1_7

learning, focuses on questions of knowledge construction, and the teacher's role is mainly one of support, stimulation, advice, and co-design.

The debate over whether direct instruction is superior to inquiry learning and vice versa has been ongoing and has recently resurfaced (see also Sweller et al. 2024; Tobias and Duffy 2009). Over time, these positions have converged, with both approaches now seen as incorporating elements of the other. Direct instruction includes not only information delivery but also problem-solving and project work. Similarly, inquiry learning is not purely autonomous; it involves teacher guidance and can include phases of explicit knowledge instruction.

The key difference between the two lies in the role and timing of problem-solving in the learning process. In direct instruction, problem situations are introduced after learners have acquired the necessary knowledge and skills. In inquiry learning, in contrast, learners develop knowledge and skills while actively working through the problem. Inquiry learning typically begins with a problem or application scenario rather than an extended phase of knowledge acquisition, as in direct instruction (see Fig. 7.1).

Both approaches can incorporate elements of cooperation and communication, but inquiry learning often encourages more group-based, cooperative learning. During the problem-solving process, exchanging ideas becomes crucial as learners engage in discussions, confront differing viewpoints, and develop new connections between concepts, which helps them refine and clarify their own understanding.

As noted earlier, neither direct instruction nor inquiry learning is inherently superior for designing learning environments. However, certain conditions may favour one approach over the other. Direct instruction is particularly well suited for

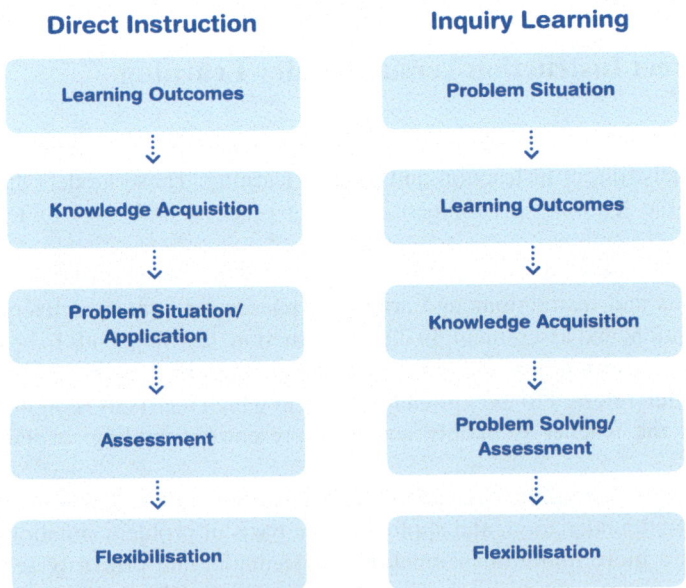

Fig. 7.1 Direct instruction versus inquiry learning

Table 7.1 Characteristics of teaching strategies

	Sequence learning activities	Significance problem situation	Learning space	Social learning
Direct instruction	★	*	*	*
Sandwich principle	★	*	*	★
Problem-based learning	*	★	*	*
Project-based learning	*	★	*	*
Inquiry learning	*	★	*	*
4C/ID	★	★	*	*
Flipped classroom	*	*	★	*
Online collaborative learning	*	*	*	★
Jigsaw	*	*	*	★

content with a hierarchical structure, where new learning builds on prior knowledge. It is also efficient for teaching and reinforcing factual information, such as safety regulations.

Conversely, inquiry learning is better suited for learners who already have some foundational knowledge, possess high intrinsic motivation, and are accustomed to independent learning. Additionally, because it demands more self-regulation and collaboration, inquiry learning supports the development of not only professional competencies but also key generic skills, such as methodological, social, and personal competencies.

Beyond the sequence of learning activities or the role of problem-solving, *learning spaces* and *social learning* can also define teaching strategies. For instance, in a flipped classroom, the learning space (whether on-site or online) shapes specific activities. In approaches such as online collaborative learning (OCL) or jigsaw, the focus is on social learning, in which interaction in various peer groups plays a critical role (see Table 7.1).

The following sections provide an overview of the teaching strategies in Table 7.1 and examine their suitability for digital learning.

7.2 Direct Instruction

Direct instruction is a structured teaching approach in which the teacher guides learners through coordinated phases of learning, continuously monitoring progress and adjusting the lesson as needed (Rosenshine and Stevens 1986). The process begins by activating prior knowledge and clearly stating the learning objectives and success criteria (see Fig. 7.2). The content is then presented, and the completion of the learning tasks is modelled. Next, learners engage in intensive practice—either individually or in groups—under supervision, aimed at promoting the transfer of knowledge to new contexts. Throughout, progress is regularly assessed, with feedback and scaffolding provided to support the learning process. It is also important that the learning content is revisited and applied at regular intervals (Retrieval Practice; see, e.g., Roediger and Butler 2011).

Fig. 7.2 Direct instruction process

Direct instruction is sometimes mistakenly equated with lecturing. While lecturing typically focuses only on content delivery, direct instruction includes additional steps, such as modelling, practice in different social constellations, and ongoing assessment with feedback.

Sandwich Principle

Another teaching strategy that can be categorized as direct instruction is the *sandwich principle* (see Fig. 7.3). Given the limitations of working memory and attention span, content delivery methods, such as presentations or instructional videos, should be kept brief (e.g., Bligh 2000) and alternated with active engagement phases. Wahl's (2013) sandwich principle addresses this by alternating between receptive phases, where content is delivered collectively, and subjective application phases, where learners practice, repeat, or apply the material. The "entrance" phase provides an overview of the content and learning objectives, while the "exit" phase offers a summary, revisits the objectives, or provides a forward-looking perspective, hence the "sandwich" analogy. The number and duration of the collective and subjective learning phases can be tailored to a specific learning context. Crucially, transitions between these phases should be well organized, with clear instructions and detailed assignments to ensure that all learners actively engage with the content.

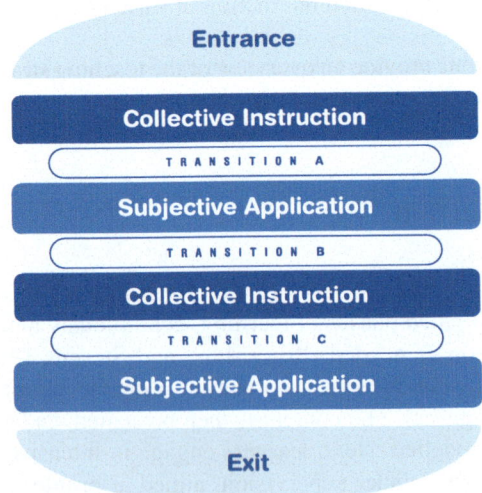

Fig. 7.3 Components of the sandwich principle

7.3 Inquiry Learning

Inquiry learning focuses on approaches in which a problem or application situation initiates the learning process, requiring learners to explore, understand, and solve it. These scenarios are designed to closely resemble real-world problems from future or current professional practice or everyday life, ensuring that they are as authentic as possible. As in real life, these problems are often complex and may include irrelevant or even contradictory information.

Problem situations play a crucial role in the learning process by creating a gap between learners' prior knowledge and the knowledge needed to solve the problem, directing the learning journey. This approach also boosts motivation by addressing issues relevant to future studies, work, or daily life (Müller 2007). A problem corresponds to a situation in which a person does not have the means and procedures to transform the initial state into the desired target state (Dörner 1987). Therefore, a well-defined problem situation should not be immediately solvable and may involve initial failure, which can be used productively in the learning process (Kapur 2008). Unlike problems, tasks are mental challenges for which the necessary methods and processes are already known.

When designing problem situations, it is crucial to focus on what the problem demands. The main problem types are outlined in Table 7.2.

Different types of problems and challenges can be assigned to different approaches to inquiry learning.

Table 7.2 Types of problems in inquiry learning

Type of problem	Challenge	Example	Request
1. Explanation	Facts or phenomena need to be explained	A company publishes its annual financial statements with a considerable loss and announces major job cuts. On the same day, the share price rises by 5%	Explain
2. Diagnosis	A deviation from the target state is detected and needs to be corrected	In a freely accessible and usable body of water, the fish population has declined sharply (tragedy of the commons)	Correct
3. Decision	An option must be chosen from alternatives (may be a moral dilemma)	Company X must decide on one of three possible locations	Decide
4. Strategy	Starting from an actual state, vaguely defined goals need to be achieved	As a central banker, you will be in charge of your country's monetary policy for the next 4 years. We are currently in a boom phase	Navigate
5. Design	An innovative creation is required from an open actual state	You become self-employed in the field of X. The banks you approach for financing require a business plan	Create

Problem-Based Learning

Problems of types 1–3 are typically addressed through *problem-based learning* (PBL), which follows an iterative learning cycle. At the start, learners are presented with a problem scenario. They begin by analysing the problem and identifying relevant facts, which helps them understand the scenario. Once the problem is clear, learners develop hypotheses for possible solutions. A key part of this step is identifying gaps in their knowledge related to the problem. At this stage, learners often work in small groups, guided by a tutor. These knowledge gaps then define the learning objectives for the self-study phase. After individually addressing these gaps, learners apply their new knowledge and evaluate their hypotheses. Finally, learners reflect on the theoretical knowledge gained throughout the process. This cycle is the basis for the widely used *"7-Step"* (or "Seven-Jump") strategy, a general process learners use to tackle problems and build knowledge. Although many PBL implementations emphasize the seven steps, their interpretation and execution can vary. The most commonly practiced version comes from Maastricht University (Schmidt 1983). Figure 7.4 provides an overview of the steps and their objectives. Notably, learners work in small tutor-supervised groups in every step except the sixth step.

Case-based learning also involves problem scenarios but is generally used to apply and practice previously learned material rather than to learn new content. This approach is closer to direct instruction.

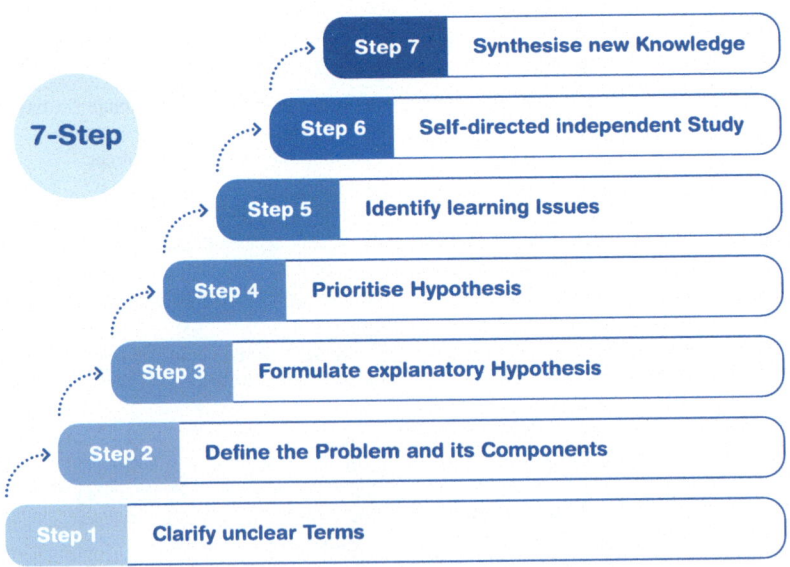

Fig. 7.4 The 7-step strategy in problem-based learning

Strategy problems (type 4) are typically addressed using *serious games* or *simulations*. These aim to model decision-making processes and the planning involved. Simulations generally have two components: the players (usually working in groups), who make decisions, and the reaction system, which processes those decisions according to specific rules. The simulation consists of the game, in which participants act and decide, and the model, which defines the framework and background of the scenario. Unlike role-playing games, in which players react to each other's actions, simulations require participants to respond to a scenario based on the predefined rules of the game model.

Project-Based Learning

Project-based learning is particularly suited for tackling design problems. In project-based learning, learners identify a problem or task, form project teams, and develop an action plan, which they execute based on self-established guidelines. This approach often involves producing creative outputs under ambiguous conditions, with multiple solutions and methods possible. Beyond acquiring relevant knowledge, project-based learning emphasizes the process of generating innovative solutions. It offers learners greater autonomy, focusing more on their active problem-solving and construction work than on direct instruction. As a result, their work is evaluated based more on its relevance to the original problem than on strict correctness.

Enquiry-Based Learning

Enquiry-based learning, also called *research-oriented* or *research-based learning*, falls under the broader category of inquiry learning. Instead of being presented with a problem, as in problem- or project-based learning, learners are encouraged to identify research gaps, formulate their own scientific questions, and independently explore answers. Throughout the research cycle, learners engage in discussions, testing, and reflection on various research methods. The focus of the evaluation is on the appropriateness of the approach rather than the absolute correctness of the solutions. Enquiry-based learning is a challenging teaching strategy, as it shares with project-based learning a high degree of learner autonomy.

Online inquiry learning was initially met with scepticism, as these strategies depend heavily on direct interactions between learners and teachers. However, with the advancement of synchronous communication and collaboration tools, which now allow for flexible social settings (e.g., breakout rooms), it has become clear—especially during the pandemic—that teaching strategies based on social learning can be effectively adapted to an online environment.

7.4   4C/ID Model

A digital learning environment does not need to adhere strictly to one of the two educational models; it can integrate elements from both direct instruction and inquiry learning. A well-known framework that blends these approaches is

van Merriënboer's (2019) *four-component instructional design* (4C/ID) model.
This model illustrates how direct instruction and inquiry learning complement
each other in developing complex professional skills, such as air traffic control,
software development, management decision-making, or medical procedures.
Like inquiry learning, the 4C/ID model emphasizes complex real-world tasks
relevant to various professional and everyday contexts. In addition to these
learning tasks, the model of four key components (4C) of instructional design
includes the elements of supportive information, procedural information, and
part-task practice, which are more closely associated with direct instruction
(see Fig. 7.5).

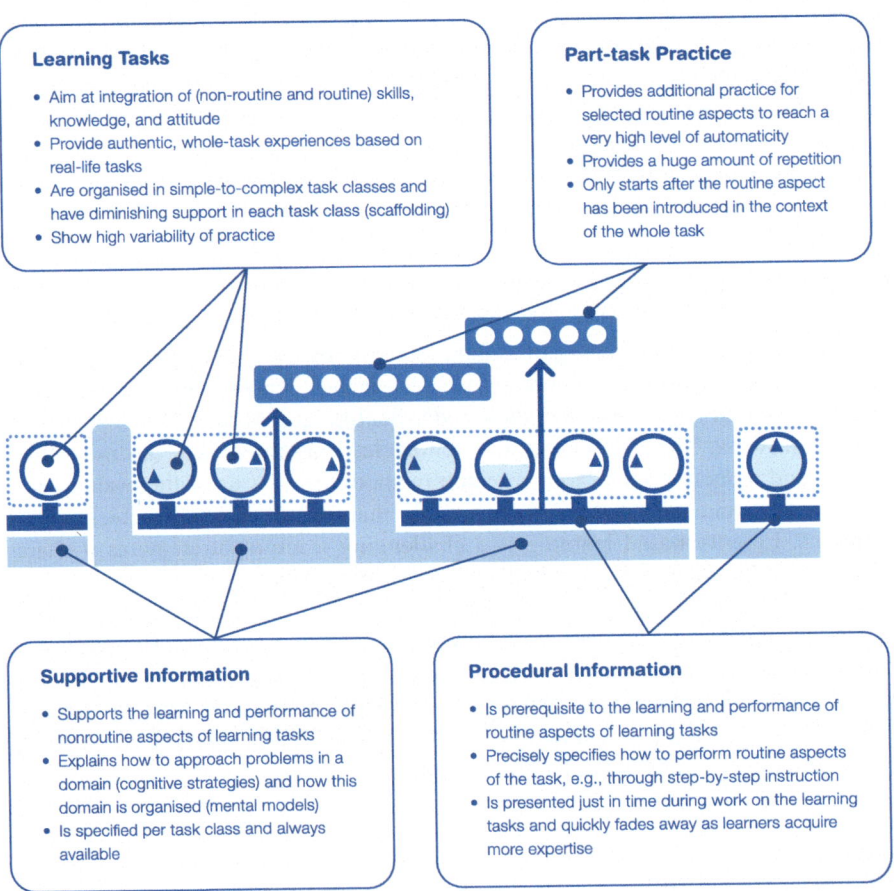

Learning Tasks

- Aim at integration of (non-routine and routine) skills, knowledge, and attitude
- Provide authentic, whole-task experiences based on real-life tasks
- Are organised in simple-to-complex task classes and have diminishing support in each task class (scaffolding)
- Show high variability of practice

Part-task Practice

- Provides additional practice for selected routine aspects to reach a very high level of automaticity
- Provides a huge amount of repetition
- Only starts after the routine aspect has been introduced in the context of the whole task

Supportive Information

- Supports the learning and performance of nonroutine aspects of learning tasks
- Explains how to approach problems in a domain (cognitive strategies) and how this domain is organised (mental models)
- Is specified per task class and always available

Procedural Information

- Is prerequisite to the learning and performance of routine aspects of learning tasks
- Precisely specifies how to perform routine aspects of the task, e.g., through step-by-step instruction
- Is presented just in time during work on the learning tasks and quickly fades away as learners acquire more expertise

Fig. 7.5 4C/ID model (van Merriënboer 2019)

To prevent cognitive overload when handling complex tasks, the 4C/ID model recommends the following strategies:

- *Increasing levels of complexity in learning tasks:* Offer various learning tasks with increasing levels of complexity, allowing learners to apply their knowledge in a wide range of contexts (e.g., electrical installations with varying complexity, such as single-way, two-way, and multiple-way switching).
- *Gradual reduction of support at the same complexity level:* Initially, learners receive substantial guidance at each complexity level. As they progress, support is gradually reduced until they can complete tasks independently. At this point, they are ready to move to the next level of complexity.
- *Tailoring supportive information to complexity levels*: Supportive information connects learners' prior knowledge with the new concepts they need to understand for nonroutine aspects of tasks. This information, often referred to as "theory," is consistent across tasks at the same complexity level and can be provided either before starting the tasks (theory first) or as needed during the tasks (theory on demand). For higher complexity levels, this information is expanded or deepened (e.g., explaining electrical circuits and circuit diagrams).
- *Just-in-time procedural information and corrective feedback*: Procedural information, such as step-by-step instructions or user manuals, supports learners in mastering the routine aspects of tasks (e.g., steps for working safely on electrical circuits and safely switching off circuit breakers). This information is provided exactly when needed and ideally paired with corrective feedback from the instructor.
- *Extensive practice of routine aspects through subtask repetition:* If mastering the routine aspects of a subtask requires high levels of automation, and regular learning tasks do not provide sufficient practice, these skills should be practiced intensively (e.g., soldering electrical installations). Wherever possible, routine practice should be embedded within broader learning tasks to help learners recognize how improving routine skills enhances their overall performance.

The design of the 4C/ID model's individual components is well documented and incorporates numerous evidence-based design principles (van Merriënboer 2019). For beginners, the media used for learning tasks can start with lower authenticity (*low fidelity*), such as written case studies or role plays, and gradually progress to more realistic contexts (*high fidelity*), eventually leading to real-world tasks (e.g., in a workplace setting). High-fidelity learning environments for tasks might include virtual production facilities, for example, using VR to create electronic installations. Drill-and-practice programmes can also be employed to practice subtasks in specific contexts, such as software training.

7.5   Collaborative Teaching Strategies

Collaborative teaching strategies foster learning processes in which the social dimensions of learning, particularly interactions among learners, are key. These interactions and communication promote sustained engagement with the content and help from a learning community where both individual and collective knowledge are developed.

According to the *community of inquiry (COI) model* (Garrison et al. 2010), the quality of collaborative learning in digital environments depends on the interaction of three key elements (see Fig. 7.6):

- *Cognitive presence:* The extent to which participants construct and share knowledge through meaningful communication.
- *Social presence:* The ability of participants to identify with the group, engage in purposeful communication, and build personal and emotional connections.
- *Teaching presence:* The effectiveness of the design and facilitation of the learning environment.

The COI model serves more as a theoretical framework than a teaching strategy, as it provides limited guidance on specific learning activities, sequences, or conditions to foster these three types of presence. However, two closely related learning strategies that offer more concrete approaches to learning activities and processes are discussed below. The online collaborative learning strategy focuses on the collaborative development of learning objectives at higher cognitive levels, such as

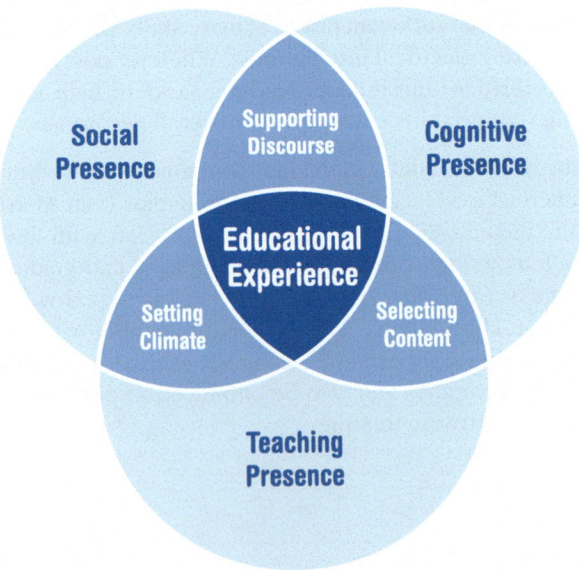

Fig. 7.6 Community of inquiry model

applying, analysing, evaluating, and creating. In contrast, the jigsaw strategy emphasizes the collaborative learning of new content at the understanding level.

Online Collaborative Learning

Online Collaborative Learning (OCL) is a teaching strategy in which knowledge is collaboratively developed by groups working on problem situations through guided online discussions (Harasim 2017). Similar to PBL, the learning process is initiated in a problem scenario. However, unlike PBL, discussions in OCL primarily occur asynchronously and in written form via forums. These forums are threaded, allowing replies to be linked to specific comments rather than displayed purely in chronological order. This structure supports the development of dynamic subtopics, with multiple replies and discussions evolving within a single thread.

Harasim (2017) outlines three key phases in the OCL process (see Fig. 7.7): idea generating, idea organization, and intellectual convergence. In the *idea generating* phase, learners contribute a range of ideas, similar to brainstorming, by collecting diverse perspectives on the topic. During the *idea organization* phase, learners analyse, compare, and categorize these ideas, engaging in activities such as agreeing or disagreeing, clarifying, elaborating, or rejecting certain views and identifying relationships between them. They may also explore new concepts by integrating external resources, such as course materials or peer suggestions. This phase helps transform individual insights into a shared understanding. Finally, the *intellectual*

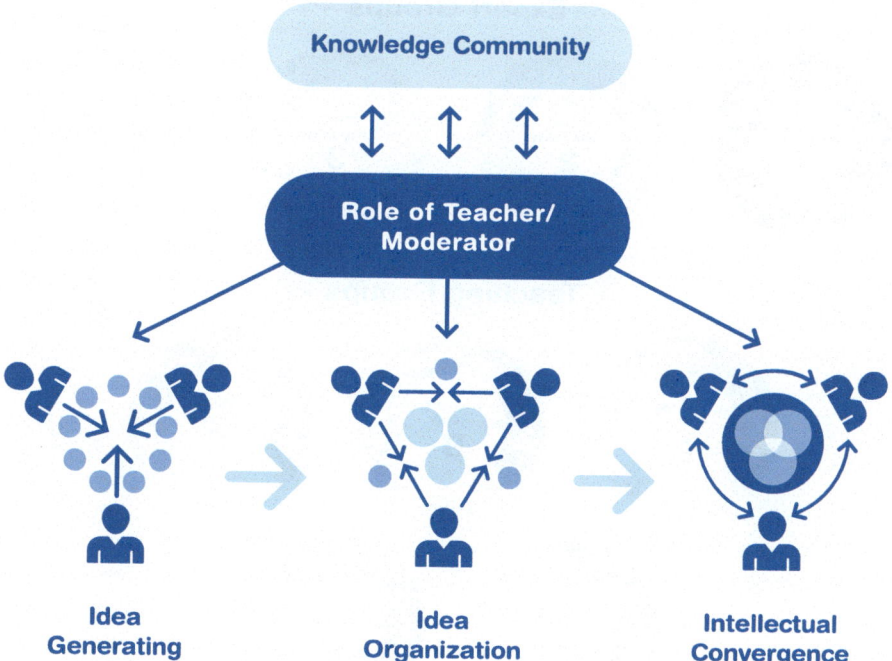

Fig. 7.7 Online collaborative learning (based on Harasim 2017)

convergence phase involves co-constructing knowledge based on this shared under-standing. Groups synthesize their ideas and viewpoints, forming collective posi-tions on the issue. The outcomes of this phase are consolidated and can be documented or presented as a final product.

OCL is not a self-directed process; teachers play a pivotal role in guiding and facilitating learning. Teachers initiate the process by presenting relevant problem situations, structuring the learning through carefully designed activities, and provid-ing ongoing support with appropriate scaffolding. They also supply additional resources and ensure that key concepts, practices, and subject standards are thor-oughly integrated throughout the learning cycle.

Jigsaw

In the *jigsaw method*, learners first work together to understand specific content and then take on the role of teachers, explaining the material to their peers. This approach is grounded in the principles of social learning and learning by teaching. In the first phase, learners are divided into *expert groups*, each focusing on a particular sub-topic or aspect of a problem. Within these groups, they review the material, discuss key questions, consider practical applications, and develop teaching aids (such as illustrations) to support their instruction.

In the second phase, the expert groups are dissolved and *teaching groups* are formed, each consisting of members from different expert groups (see Fig. 7.8). In

Fig. 7.8 Functionality of the jigsaw

this step, the "experts" teach their peers the material they have mastered, responding to questions and addressing any challenges their fellow learners might raise.

The jigsaw method can be effectively implemented online by combining asynchronous and synchronous learning phases. Expert group work typically occurs asynchronously, with synchronous meetings scheduled by the group. Teaching group meetings can be held synchronously in small breakout groups via video conferencing. The online format offers several advantages: it accommodates large numbers of learners and groups, simplifies logistical organization (such as managing rooms and technology), and allows teachers to easily join groups to offer support when needed.

 ### Tips for Teaching Strategy

- **Align with learning objectives:** Choose teaching strategies based on the objectives of the digital learning experience, whether delivered online, on-site, asynchronously, or synchronously.
- **Tailor to context:** Use the teaching strategies as models and inspiration, adapting and customizing them to suit your unique teaching context and needs.
- **Incorporate variety:** Introducing diverse teaching strategies, tools, and media enhances curiosity and motivation, so vary your educational design to keep learners engaged.

 ### Teaching Strategies in myScripting

myScripting is not tied to a specific teaching strategy but serves as a design tool to help create diverse and reflective learning environments tailored to the learning context. In myScripting, scripts for the teaching strategies discussed here are available. These can be reviewed, copied, and integrated into your own educational designs.

References

Bligh, D. A. (2000). *What's the use of lectures?* Jossey-Bass.

Dörner, D. (1987). *Problemlösen als Informationsverarbeitung* (Vol. 3). Kohlhammer.

Garrison, D. R., Anderson, T., & Archer, W. (2010). The first decade of the community of inquiry framework: A retrospective. *The Internet and Higher Education, 13*(1), 5–9. https://doi.org/10.1016/j.iheduc.2009.10.003

Harasim, L. (2017). *Learning theory and online technologies*. Routledge.

Kapur, M. (2008). *Productive failure*. *Cognition and instruction, 26*(3), 379–424. https://doi.org/10.1080/07370000802212669

Müller, C. (2007). *Implementation von Problem-based Learning: eine Evaluationsstudie an einer Höheren Fachschule*. hep-Verlag.

Roediger, H. L., & Butler, A. C. (2011). The critical role of retrieval practice in long-term retention. *Trends in cognitive sciences, 15*(1), 20–27. https://doi.org/10.1016/j.tics.2010.09.003

Rosenshine, B., & Stevens, R. (1986). Teaching functions. In M. C. Wittrock (Ed.), *Handbook of research on teaching* (3 ed., pp. 376-391). Macmillan.

Schmidt, H. G. (1983). Problem-based learning: Rationale and description. *Medical Education, 17*(1), 11–16. https://doi.org/10.1111/j.1365-2923.1983.tb01086.x

Sweller, J., Zhang, L., Ashman, G., Cobern, W., & Kirschner, P. A. (2024). Response to De Jong et al.'s (2023) paper "Let's talk evidence – The case for combining inquiry-based and direct instruction". *Educational Research Review, 42*, 100584. https://doi.org/10.1016/j.edurev.2023.100584

Tobias, S., & Duffy, T. M. (2009). *Constructivist instruction: Success or failure?* Routledge.

van Merriënboer, J. J. (2019). *The four-component instructional design model: An overview of its main design principles*. https://www.4cid.org/wp-content/uploads/2021/04/vanmerrienboer-4cid-overview-of-main-design-principles-2021.pdf

Wahl, D. (2013). *Lernumgebungen erfolgreich gestalten: Vom trägen Wissen zum kompetenten Handeln*. Julius Klinkhardt.

Chapter 8
Learning Activities

Learner engagement is crucial for effective learning, making it the teacher's responsibility to design activities that actively involve learners. In digital learning environments, key activities include practice, case work, content discussions, and assessment of learning progress. Digital tools also introduce new possibilities, such as simulations and VR, which foster innovative learning approaches. It is important that learning activities not only replicate conventional teaching but also expand and transform teaching practices (Kimmons et al. 2020).

This chapter explores the following key questions: What types of learning activities can be distinguished? How do these activities contribute to learning success? How should learning activities and their related media be effectively designed?

8.1 Content Delivery

Content can be delivered in a variety of formats, including text, images, animation, and video. Table 8.1 shows an overview of the specific capabilities of each medium for *content delivery* to help make a choice.

It is crucial to present content clearly and understandably, especially when dealing with complex materials. Building on similar assumptions about human information processing as cognitive load theory (see Sect. 2.1), Richard E. Mayer developed the *cognitive theory of multimedia learning* (CTML, Mayer 2002, 2020). According to CTML, working memory has limitations, and people process information through two distinct channels: visual/pictorial and auditory/verbal. This leads to the *multimedia principle*, which suggests that learning is more effective when information is presented with instructional text and relevant visuals.

© The Author(s) 2025
C. Müller, *Digital Learning Design*, SpringerBriefs in Education,
https://doi.org/10.1007/978-3-031-89045-1_8

Table 8.1 Media for content delivery

Medium	Opportunities
Animation, simulation, and virtual reality	Visualizing processes and workflows; practicing interaction with simulated objects and machines
Lecture recording	Accessible anywhere, anytime; adjustable playback speed (e.g., fast-forward and rewind)
Video	Offers insights into authentic contexts, difficult-to-access information, historical documents, and contemporary testimonies
Physical textbook	Provides a tactile experience; usable without internet access
Online textbook	Integrates multimedia and interactive elements; easily updated; cost-effective distribution
Podcast	Available anywhere, anytime (e.g., during activities like jogging); simpler production than video

Mayer also introduced in the CTML a series of additional principles that serve as guidelines for designing media in digital learning environments (see Table 8.2).

To ensure that digital learning environments are accessible to individuals with disabilities, it is important to follow national and international accessibility standards during media production. The following simple, effective measures can enhance *accessibility*.

Table 8.2 Principles of multimedia learning (Mayer 2020)

Principle	Description
Multimedia principle	People learn better from words and pictures than from words alone
Modality principle	People learn better from graphics and narration than from graphics and onscreen text
Redundancy principle	People do not learn better when printed text is added to graphics and narration
Spatial contiguity principle	People learn better when corresponding words and pictures are presented near rather than far from each other on the page or screen
Temporal contiguity principle	People learn better when corresponding words and pictures are presented simultaneously rather than successively
Segmenting principle	People learn better when a multimedia lesson is presented in user-paced segments rather than as a continuous unit
Coherence principle	People learn better when extraneous material (images, sounds, etc., that are irrelevant to the comprehension process) is excluded rather than included

 **Tips for Accessibility**

Structure and Navigation

- **Consistency:** Maintain a consistent structure and format for all learning activities.
- **Clear structure:** Use headings, paragraphs, lists, and tables to organize content for easier navigation and understanding.
- **Simple, consistent navigation:** Ensure that navigation on websites or learning platforms is straightforward and uniform.
- **Use anchors:** Include "Skip to Content" links and anchors to help learners quickly reach key sections.

Text and readability

- **Clear, simple language:** Write in easy-to-understand language with clear sentence structures and define technical terms.
- **Alternative text for multimedia:** Provide alternative text for images, videos, and multimedia to support learners with visual or hearing impairments.
- **Readable fonts and contrasts:** Use easy-to-read fonts (e.g., sans-serif) and ensure strong contrast between text and background.

Multimedia content

- **Subtitles and transcripts:** Include subtitles and transcripts for all video and audio content to ensure accessibility for deaf or hard-of-hearing learners.
- **Audio description:** Add audio descriptions for important visual elements.
- **Flexible media options:** Offer content in multiple formats (e.g., text, audio, and video) to cater to diverse learning needs.

Compatibility with assistive technologies

- **Screen reader compatibility:** Ensure all content works with screen readers by adhering to HTML standards and Accessible Rich Internet Applications guidelines.
- **Keyboard navigation:** Make all functions and content accessible via keyboard-only navigation without requiring a mouse.
- **Accessibility testing:** Regularly test websites (e.g., using tools like WAVE) or documents (Word/PDF) to ensure compliance with accessibility standards.
- **User feedback:** Gather input from users with disabilities to assess real-world accessibility and user friendliness.

Conformity

- **Standards and guidelines:** Follow Web Content Accessibility Guidelines and local accessibility laws.
- **Cultural sensitivity:** Avoid language or tasks that may alienate or demotivate learners from diverse cultural backgrounds.

Educational videos effectively combine visual and auditory channels, making them ideal for conveying information. Their popularity has grown due to their wide distribution on video platforms and social media. Learning videos can be classified into informational videos and explanatory videos.

Informational videos are used to capture attention, introduce content, and motivate the learning process. These short videos often serve as teasers for courses or content blocks, using multimedia elements such as images, videos, and music to engage learners.

Explanatory videos, such as those from Khan Academy, have significantly influenced digital learning. Research (Noetel et al. 2021) has shown that these videos can enhance learning in various ways:

- Dynamically visualizing abstract or complex concepts (e.g., physical processes or temporal compression of biological processes)
- Providing realistic experiences from professional fields or hard-to-access environments (e.g., demonstrations of industrial production or field experiments)
- Demonstrating practical skills in professional settings, sports, or social behaviour (e.g., illustrating movements in physiotherapy or providing step-by-step software tutorials)

Explanatory videos are effective because they engage both visual and auditory channels. However, optimizing cognitive load is critical for maximizing learning. The following *guidelines for video production*, many based on multimedia principles (*in italics*), can improve their effectiveness (Guo et al. 2014; Mayer 2021):

Structuring the video

- Break videos into segments of 3–9 min (*segmentation principle*) and allow learners to control the pace (*learner control*).
- Follow each section with a learning activity, such as quizzes (generative activities; see also Sect. 8.2).

Designing image and sound

- Do not just rely on text but also use meaningful images so that learners can make the best use of both the visual and auditory systems (*multimedia principle*).
- When meaningful images are not an option, use some visually represented terms (e.g., on slides) so that learners can use both the visual and auditory systems in some way (*modality principle*).
- If graphics and illustrations are presented with a verbal description, the simultaneous text (labels, closed captioning, etc., but not subtitles for foreign language speakers) should be removed (*redundancy principle*).
- Present related words and images as closely together in time and space as possible so that learners do not have to use their working memory to link them (*spatial and temporal contiguity principle*).
- Remove words or images that are not directly related to the main learning objectives so that learners do not have to use cognitive resources to process irrelevant content. Ensure that any additions (e.g., colour, animation, or human features) are meaningful and not distracting (*coherence principle*).
- Where possible, animate visual material in a meaningful way so that learners can understand how something moves and how it works.

- Make abstract concepts tangible by giving them agency and human characteristics (for example, an "aggressive" virus that "attacks" the immune system drawn with a sinister look).
- Make the images visually appealing by using pleasant colours in a targeted way, for example, by using a contrasting colour to emphasize something important (*emotional design*).

Behaviour/language teachers:

- Include a person/figure on the screen only if it helps to draw attention to the essential (*image principle*).
- Use simple and personal language in the I/we form and link the learning content to the learners' experiences so that they can build on their previous knowledge; bring in the personality of the teacher with preferences, experiences, etc. (*personalization principle*).
- Highlight key content through verbal emphasis, pointers, arrows, etc., so that learners focus on it (*signalling principle*).
- Use dynamic gestures and movements, and maintain eye contact (*embodiment principle*).
- Film the actions of a demonstration from a first-person perspective (*perspective principle*).

The cost of producing educational videos should not be underestimated, and the use of explanatory videos in educational design should be carefully considered in a *cost–benefit analysis*. While it is relatively easy today to produce explanatory videos using screen casting technology, recording a practical demonstration requires more experience and technical equipment. Additionally, maintaining media can be challenging, so sustainability should be considered. Finally, before producing a new explanatory video, explore existing resources that may be suitable for the purpose, reducing the need for in-house production.

 Tips for Content Delivery

- **Media mix:** Use a variety of media to deliver content, such as instructional texts, graphics, animations, and videos.
- **Cost–benefit analysis of video production:** While videos can enhance learning, they can be costly to produce. Conduct a thorough cost–benefit analysis for in-house production, and consider utilizing open educational resources (OER).
- **Adapt cognitive load:** Ensure that the cognitive load of media is appropriate for learners' competence levels.
- **Use of artificial intelligence (AI):** Leverage AI tools to support the creation of learning materials (e.g., instructional texts, videos, graphics, and animations) and the development of tasks and assessments.
- **Copyright awareness:** Be mindful of copyright considerations when designing learning materials.

8.2   Activation

The teacher's role is to design learning activities that are purposeful and engaging, encouraging learners to actively participate and achieve learning objectives. Learning resources can prompt different *levels of engagement* (see Table 8.3). For example, learners might passively watch an explanatory video or skim a text, which results in *passive* learning engagement. However, when learners pause, rewind, or take notes, this reflects *active* learning engagement. If learners connect the content to prior knowledge or explain it in more detail, they demonstrate *constructive* learning engagement. The highest level, *interactive* learning engagement, occurs when learners discuss content with peers or explain tasks to others.

The *ICAP model* (Chi and Wylie 2014) outlines these different levels of engagement and suggests that deeper interaction with and discussion of the content leads to more effective learning outcomes.

Learning engagement can be enhanced through well-designed *learning tasks* that guide learners in how to effectively use learning resources. Many learners, especially in virtual environments, lack experience with online interaction and require specific guidance on organizing and facilitating collaboration (Vogel et al. 2017). Effective learning tasks include contributing to forum discussions (with contributions and feedback), collaboratively developing texts or videos using annotation tools, creating content through a wiki, explaining material to peers (e.g., through jigsaw activities), and providing feedback on each other's work or projects.

Traditional *learning strategies*, such as underlining and highlighting, have been shown to be less effective (Dunlosky et al. 2013). More effective approaches involve learners summarizing in their own words, explaining concepts to others,

Table 8.3   Learning engagement according to the ICAP model (Chi and Wylie 2014)

Principle	PASSIVE Receiving	ACTIVE Manipulating	CONSTRUCTIVE Generating	INTERACTIVE Dialoguing
Lecture	Listening without doing anything else but oriented towards instruction	Repeating or rehearsing; copying solution steps; taking verbatim notes	Reflecting out loud; drawing concept maps; asking questions	Defending and arguing a position in dyads or small groups
Text	Reading entire text passages silently/aloud without doing anything else	Underlining or highlighting; summarizing by copy-and-delete	Self-explaining; integrating across texts; taking notes in one's own words	Asking and answering comprehension questions with a partner
Video	Watching the video without doing anything else	Manipulating the video by pausing, playing, fast forwarding, rewinding	Explaining concepts in the video; comparing and contrasting with prior knowledge or other materials	Debating with a peer about the justifications; discussing similarities and differences

reproducing worked examples, or answering practice exam questions [see Fiorella and Mayer (2015) for generative learning strategies]. In adult education, reflecting on key takeaways after a topic block can support learning and transfer. For example, the *3-2-1 reflection* method asks learners to identify three key lessons, explore two questions or topics they want to know more about, and note one insight into how the content impacts their learning or life.

 Tips for Activation

- **Learning tasks:** Motivate learners to engage deeply with the content by incorporating learning tasks that promote active involvement, especially when using instructional texts or explanatory videos.
- **Variety in learning activities:** Offer a range of learning activities. For example, do not just provide multiple-choice (MC) tests at the end of the learning sequence but also provide alternative formative learning assessments and feedback (see also Assessment).
- **Generative learning strategies:** Encourage the active processing of new knowledge using generative learning strategies that promote effective learning, such as learners restating the learning content to themselves or explaining it to each other.

8.3 Interaction

Online learning differs from in-person classroom instruction, primarily due to the impact of temporal and spatial distance on *interaction*. This distance necessitates extra attention to motivation in both synchronous and asynchronous interactions—whether between teachers and learners or among learners themselves.

Synchronous interactions typically occur in webinars, which are virtual meetings or lectures. Below are some tips for designing webinars that foster motivation and support effective learning.

 **Tips for Webinars**

- **Technology**: Provide instructions on the necessary tools and equipment in advance and offer opportunities for testing.
- **Clear participation expectations**: Establish guidelines for camera/microphone use, speaking requests, nonverbal responses, and chat contributions and to maintain distraction-free real or virtual backgrounds.
- **Variety**: Break the webinar into short and varied segments and incorporate different learning activities and media.
- **Focus attention**: Capture attention with storytelling, humour, or provocative statements.
- **Visual aids**: Use a variety of visual aids, such as presentations, videos, and interactive whiteboards.
- **Interaction**: Encourage engagement through methods such as think-pair-share, buzz groups, role-playing, or group work in breakout rooms.
- **Surveys**: Use polls to gather and discuss learners' experiences and opinions.
- **Exercises and quizzes**: Include short quizzes or exercises to check learning and apply knowledge.
- **Offline time**: For extended online events (e.g., full-day sessions), incorporate offline learning time to provide breaks from continuous screen engagement.
- **Teacher presence**: Be actively present, respond quickly to learner input, and make yourself available before and after the session.

Synchronous phases in MOOCs and corporate training with a global participant base can be challenging due to time zone differences. However, in asynchronous learning environments, it becomes even more important to foster *asynchronous interaction* through methods such as forum discussions, peer feedback, or the organisation of smaller learning groups to promote a sense of community and social inclusion (see the COI model in Sect. 7.5).

Although *forums* are commonly used in online courses, many learners find them more burdensome than engaging, and teachers are often disappointed with the quality of discussions. This is usually because of poorly designed *discussion prompts*. Below are two examples of forum prompts addressing a company's compliance with data protection guidelines:

Prompt 1: "Why is it important for employees to comply with a company's privacy policies? Give two reasons."

Prompt 2: "Imagine an employee has accidentally sent sensitive customer data via email to an external partner. The incident has not been reported yet, and the employee is considering whether to conceal the mistake or inform the supervisor, knowing it could have serious consequences for both him and the company. What should the employee do? What potential impacts could this decision have on both him and the company?"

The first prompt simply asks learners to recall information, which could be better suited as a test question rather than a forum discussion. The second prompt places learners in a real-world scenario, encouraging them to consider ethical, legal, and business implications. This type of prompt fosters a richer discussion, allowing participants to share different perspectives on issues such as transparency, risk management, and the importance of trust and compliance within an organization. Prompts that ask learners to apply knowledge, reflect on their own experiences, and make decisions are far more engaging and can lead to dynamic discussions in which participants actively respond to each other's contributions.

The following tips help transform forums into dynamic and effective learning environments that promote active participation and deep engagement with course content.

 **Tips for Forums**

- **Practical relevance:** Use realistic scenarios to spark discussion and prompt learners to share practical examples, experiences, and personal insights.
- **Encourage multiple perspectives:** Formulate open-ended questions that stimulate critical thinking and allow for diverse viewpoints. Consider incorporating role-play activities where learners argue from different perspectives.
- **Foster interaction:** Motivate learners to respond to each other's posts to build a sense of community.
- **Reward participation:** Assign points or grades for participation or link discussions with tasks like summaries or reflections to emphasize relevance and deepen involvement.
- **Netiquette:** Set clear guidelines for respectful, constructive discussions, and provide examples when necessary.

While online media are widely used, online learning remains unfamiliar to many learners, requiring proper introduction and support. *Salmon's five-stage model* (2013) offers a structured framework to guide and develop learners in an online environment (see Fig. 8.1).

Stage 1—*Access and motivation*: In this initial stage, learners are introduced to the online learning platform and welcomed (e.g., through a welcome video). It is crucial to encourage learners to access content, engage with peers (e.g., in "meet & greet" forums), and reassure them that technical support is available if needed.

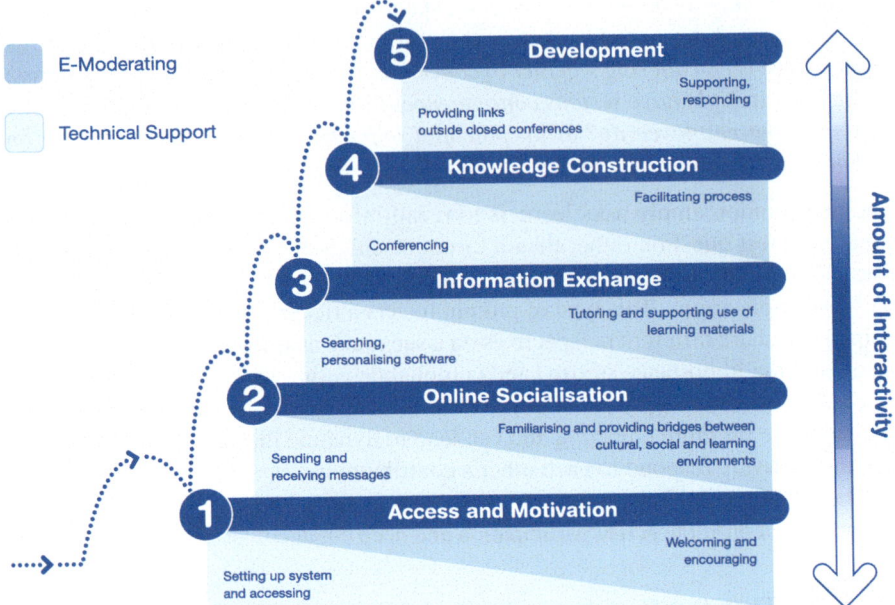

Fig. 8.1   Five-stage model according to Salmon (2013)

Stage 2—*Online socialization:* Learners begin to expand their networks and communicate within the online community. Moderated discussions ensure appropriate interactions. An effective activity could be an introduction forum where learners share their experiences or goals (e.g., "What are my experiences with...?" or "What are my goals?").

Stage 3—*Information exchange*: In this stage, learners actively engage with learning materials and organize their learning processes. Teachers should provide structured tasks, clear instructions, and communication tools while scheduling the learning process effectively and clarifying how the results will be presented.

Stage 4—*Knowledge construction*: At this point, learners should take more control over their learning. Teachers should monitor progress and offer targeted support, as online learning can sometimes feel isolating and learners may struggle without proper guidance.

Stage 5—*Development:* Learners at this stage become confident and independent, able to lead group discussions and apply their knowledge to real-world contexts. Teachers should offer additional learning resources to support further development and reflection on their competencies.

Each stage requires learners to master specific technical skills (shown at the bottom left of each stage in Fig. 8.1) and develop e-moderation skills (displayed at the top right). The interactivity bar on the right side illustrates the increasing intensity of interactions, starting with minimal engagement in Stage 1 and peaking around

Stage 3. By Stage 5, learners typically return to more individual tasks. At the knowledge construction level, the aim is not just to exchange information but to collaboratively build knowledge.

The use of *online collaboration* in the learning process requires careful planning. While substantial evidence supports the effectiveness of collaborative learning, it can be particularly beneficial by extending individual cognitive capacity. By "pooling" the working memory of team members, learners can tackle more complex tasks (Kirschner et al. 2011), Additionally, online collaboration can reduce the extrinsic load by streamlining group organization. For example, small-group synchronous discussions can be efficiently managed in breakout rooms, where learners can exchange ideas quickly while still accessing additional resources or teacher support. However, technical issues and a lack of social cohesion can increase the extrinsic load in online learning. Collaborative learning introduces an extrinsic load due to the need to organize and manage interactions (Janssen et al. 2021). This can lead to disagreements among learners and is a common criticism of group work. To ensure that the benefits of online collaboration outweigh the challenges, careful design is essential.

A key element of successful collaboration is *positive interdependence*—learners must understand that their goals and success are interconnected. Learning tasks should be structured to highlight this interdependence, ensuring that each learner knows the group can only succeed if they all work together, as Johnson et al. (2008) describe: "Group members need to know that they sink or swim together." This can be achieved by assigning challenging tasks that cannot be completed individually, with each learner contributing the unique skills and resources necessary for the group's success. Additionally, assigning specific roles, such as facilitator, timekeeper, recorder, or summarizer, ensures that each group member has clear responsibilities and that the group functions efficiently.

Group composition also influences the effectiveness of collaboration. Larger groups offer more perspectives and resources but come with higher transaction costs and lower social cohesion, which can lead to dysfunctional dynamics like the free-rider effect. For complex tasks, groups of 4–5 members are ideal, while simpler tasks can be handled by groups of 2–3. In forming groups, prior knowledge should also be considered. Clark and Mayer (2023) recommend homogeneous teams for simpler tasks and heterogeneous teams for more complex problems.

Online collaboration can be further enhanced through *collaboration scripts* (Vogel et al. 2017), which provide structured guidance on how learners should interact and work together effectively. These scripts outline process steps, allocate roles and tasks, provide timeframes, and offer templates and tools to support collaboration (Kollar et al. 2006).

Finally, the *social presence* of the teacher plays a central role in fostering interaction. Frequent communication, timely feedback, and the sharing of personal values and interests can build loyalty to the teacher and model effective interaction for learners.

Tips for Interaction

- **Clear structure and communication:** Structure and organize the course well, and communicate clearly and repeatedly through different channels.
- **Teacher presence:** Maintain a strong presence, especially early in the course, by responding quickly to questions and engaging regularly with learners.
- **Encouraging learner interaction:** Foster asynchronous collaboration through forums, peer feedback, and group assignments.
- **Engaging webinars:** Design interactive synchronous sessions with tools such as chats, whiteboards, polls, and breakout rooms for varied social learning formats.
- **Supporting collaboration:** Implement learning assignments with positive interdependence, organize groups thoughtfully, and use collaboration scripts to guide teamwork.

8.4   Assessment

Assessment refers to the system of evaluation, assessment, and feedback, serving several key functions. It shapes learners' perceptions of teaching and learning, generates insights into teaching quality, and ensures the effectiveness of learning environments (assurance of learning). Crucially, assessment also strongly guides how learners plan their learning. For example, if multiple-choice (MC) questions focus primarily on factual knowledge, learners may engage in surface-level learning. However, MC questions can also assess higher-order thinking skills—such as understanding, applying, and analysing—when designed thoughtfully, such as scenario-based questions (see also Di Giusto et al. 2019). Following the principle of *constructive alignment* (Biggs 1999), assessments should align with learning objectives and the learning environment.

Research highlights the importance of *feedback* for learning (Black and Wiliam 2010; Hattie 2009). Feedback in education refers to the response to a learner's answer, task, or performance. Ramsden (1991) found that high-quality feedback, which is timely, detailed, and focused on performance, is particularly conducive to learning. Feedback should address three critical questions (Hattie and Timperley 2007): it is important to have clearly defined goals ("Where do I need to go?"), information about current performance in relation to those goals ("What path am I on?"), and guidance that the learner can use in the further learning process ("What do I need to do next?"). Effective feedback not only identifies errors but also provides specific advice for improvement. However, the usefulness of feedback depends on how learners engage with it. If feedback is accompanied by a grade, the focus may shift away from the feedback itself, and a negative grade can be demotivating. This limits the effectiveness of feedback in summative assessments, in which learners may not immediately benefit from the suggestions provided.

   In online and blended learning, feedback from teachers is also crucial for addressing individual learning needs and preventing social isolation. Feedback on online contributions should be personalized to show that learners' efforts are recognized and valued rather than using generic comments. Frequent personalized feedback from teachers and peers fosters a sense of belonging to a learning community, increasing learner engagement. Teachers understand the importance of feedback for learning, but providing individual feedback—such as on learning tasks or formative tests—can be very time-consuming. As a result, teachers with limited time resources often reduce formative assessments first. To balance the need for personalized feedback with teachers' limited time, the following strategies can be helpful:

- *Electronic tests with feedback:* These tests provide learners with immediate feedback on their progress and tips on correct answers (e.g., for MC questions). Integrated AI can provide feedback on open-ended questions.
- *Self-assessment:* Learners can assess and reflect on their progress using predefined criteria, standards, best-practice examples, and reference solutions.
- *Peer feedback:* Learners receive feedback from their peers, with tools automating the process, including criteria and standards for assessment.
- *Focus on formative assessment:* Implement a two-stage process where feedback (without grades) is provided during the learning process, and summative assessment (with only grades) occurs at the end.
- *Limit feedback to a few key points:* Learners are unlikely to process extensive feedback. Three well-considered personalized comments are good guidelines for feedback.

LMSs increasingly offer *learning analytics* that provide real-time feedback based on learners' interactions with the system, helping them optimize their learning process (Ifenthaler and Yau 2020).

 Tips for Assessment

- **Congruence:** Ensure that assessment aligns with learning outcomes and the overall learning environment (constructive alignment).
- **Guidance:** Use assessments to direct learners' attention to key content and processes and to reward engagement.
- **Diverse methods:** Employ a range of assessment methods (e.g., written exams, oral exams, reports, and portfolios) to capture learning outcomes more validly.
- **Feedback:** Offer frequent, timely, and personalized feedback on both learning outcomes and the learning process.
- **Electronic support:** Incorporate automated feedback systems to provide insights into correct solutions or solution paths during electronic assessments.
- **Workload:** Focus personalised feedback on formative assessments and limit feedback to a few targeted comments to balance learner needs with teacher time constraints.

8.5   Learning Assignments

Learning assignments integrate key elements of content delivery, activation, inter-action, and assessment, guiding learners through the digital learning environment. To be effective, assignments should lead learners to the learning objectives effi-ciently, with minimal effort, while maintaining their interest.

Digital learning offers new possibilities for differentiation and personalization. Adaptive learning systems can analyse learners' progress, provide targeted feed-back, and offer *personalized learning*. Figure 8.2 illustrates different learning paths for learners with varying prior knowledge (light blue for low and dark blue for high).

Adaptive learning environments have long been achievable in LMSs, such as Moodle, in which activity completion settings allow specific activities to be unlocked based on learner progress. Similarly, *branching scenarios* in authoring tools enable the creation of interactive stories that let learners make decisions, actively shaping their learning paths. With the integration of AI, the ability to analyse each learner's progress, strengths, and gaps has not only expanded signifi-cantly but has also accelerated, making it possible to personalize content and assessments more quickly and effectively to align with individual goals, skills, and preferences.

Below are essential guidelines for designing effective learning assignments.

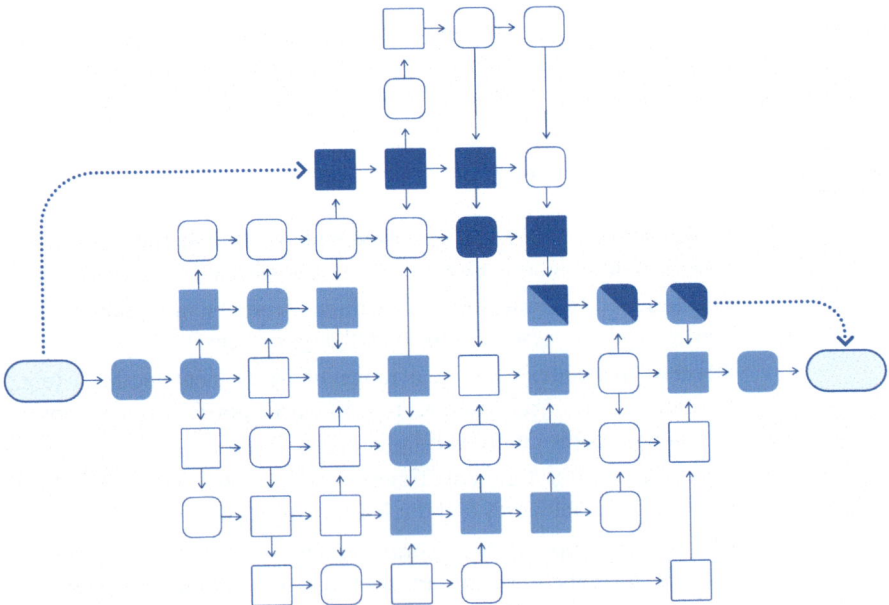

Fig. 8.2 Personalized learning paths using adaptive learning systems

 **Tips for Learning Assignments**

- **Relevance and benefit:** Clearly explain the importance of each assignment and how it contributes to the learning objectives.

- **Connection to learners' lives:** Design assignments that relate to learners' personal or professional experiences. Use real-world scenarios and motivating language to boost engagement and encourage the application of learned concepts.

- **Appropriate complexity:** Ensure assignments are appropriate in terms of scope and difficulty, providing appropriate support. Break down overly complex tasks into smaller related subtasks or provide support, such as worked examples, to help learners process the material.

- **Clear instructions:** Provide precise instructions, including required steps, resources, time expectations, deadlines, and available support.

- **Transparent assessment:** Clearly communicate the expected outcomes, how they will be assessed, and the criteria for evaluation. Specify how and when feedback will be provided (e.g., self-assessment, peer, automated, or teacher feedback).

- **Personalization:** Offer flexible learning paths and tasks that cater to individual knowledge levels and interests. Provide additional challenges for advanced learners and support for those who need extra help.

This chapter concludes by examining the *relationship between the different learning activities*. While prioritizing activity, interaction, and assessment over content delivery is generally effective, the optimal balance depends on the specific learning context. For straightforward objectives, such as teaching basic rules or procedures, a course combining content delivery through text, images, multimedia, and a final comprehension test may be sufficient. In contrast, developing advanced skills at higher cognitive levels requires a learning environment focused on active engagement, meaningful interaction, and frequent assessment with feedback, applied within diverse authentic contexts.

myScripting   **Learning Activities in myScripting**

The educational design is carried out in myScripting in the designer view. The *topics* are arranged vertically and the script is structured in terms of content, while the *learning phases* are arranged horizontally and structured in terms of time. For educational design, teachers select from predefined or self-defined learning activities and place them in sequence. Information on the educational function and technical implementation is provided for each learning activity.

Since myScripting is tailored to digital learning environments, its *learning activities* are similar to those found in various LMSs. For example, the Moodle activity set includes activities available in the Moodle LMS. If no specific LMS is used, the "Other" activity set can be selected, containing a range of activities from common LMS platforms.

Learning activities are categorized into four groups—*content delivery*, *activation*, *interaction*, and *assessment*—based on their primary educational function. However, some activities can overlap categories. For instance, a forum can be used for activation, interaction, and even assessment. To distinguish these, content delivery activities are coloured dark blue, while activation, interaction, and assessment activities are coloured light blue.

As discussed in Section 8.2, *learning assignments* can enhance active engagement with content. In myScripting, assignments can be tailored for each activity and aligned with an ICAP level. Each ICAP level comes with suggestions for assignments. Formulating engaging learning assignments is particularly important for content delivery activities, such as instructional videos or lectures, while activation, interaction, and assessment activities inherently promote engagement. The example below illustrates the textbook activity, offering one passive, three active, three constructive, and one interactive assignment, which can also be combined.

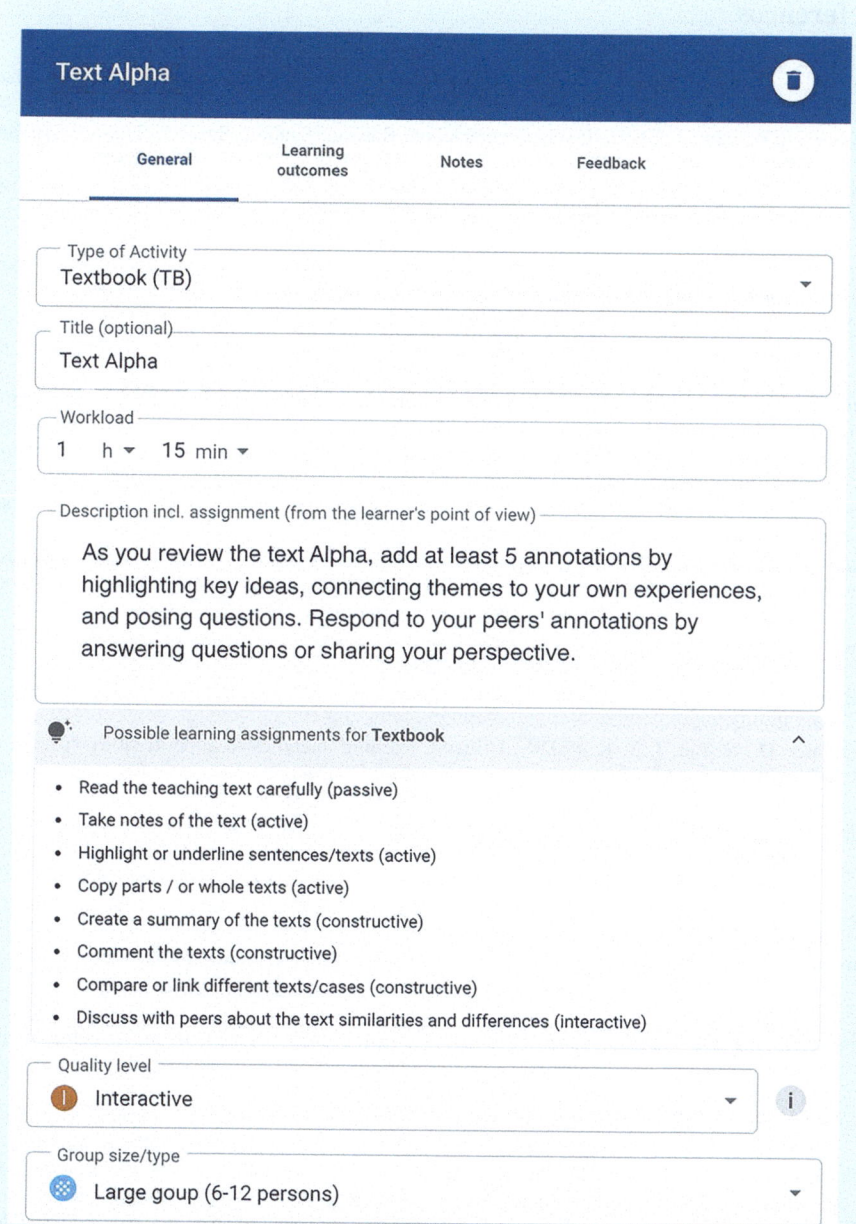

On the designer view, learning assignments are labelled with specific buttons for each ICAP level, and the number and distribution of these levels are displayed in the analytics view. This enables teachers to effectively plan and evaluate their learning activities.

References

Biggs, J. B. (1999). *Teaching for Quality Learning in University*. Society for Research in Higher Education and Open University Press.

Black, P., & Wiliam, D. (2010). Inside the black box: Raising standards through classroom assessment. *Phi Delta Kappan, 92*(1), 81–90. https://doi.org/10.1177/003172171009200119

Chi, M. T., & Wylie, R. (2014). The ICAP framework: Linking cognitive engagement to active learning outcomes. *Educational psychologist, 49*(4), 219–243. https://doi.org/10.1080/0046152 0.2014.965823

Clark, R. C., & Mayer, R. E. (2023). *E-learning and the science of instruction: Proven guidelines for consumers and designers of multimedia learning*. Wiley.

Di Giusto, F., Müller, C., Reichmuth, A., Adams-Hausheer, D., & Christian, J. (2019). *Multiple-choice questions: Teaching guide for higher and professional education*. Zurich University of Applied Sciences. https://digitalcollection.zhaw.ch/handle/11475/19339

Dunlosky, J., Rawson, K. A., Marsh, E. J., Nathan, M. J., & Willingham, D. T. (2013). Improving students' learning with effective learning techniques: Promising directions from cognitive and educational psychology. *Psychological Science in the Public Interest, 14*(1), 4–58. https://doi.org/10.1177/1529100612453266

Fiorella, L., & Mayer, R. E. (2015). *Learning as a generative activity*. Cambridge University Press. https://doi.org/10.1017/CBO9781107707085

Guo, P. J., Kim, J., & Rubin, R. (2014). How video production affects student engagement: An empirical study of MOOC videos. *Proceedings of the first ACM conference on Learning @ scale conference*, Atlanta, Georgia, USA. https://doi.org/10.1145/2556325.2566239

Hattie, J., & Timperley, H. (2007). The power of feedback. *Review of Educational Research, 77*(1), 81–112. https://doi.org/10.3102/003465430298487

Hattie, J. A. C. (2009). *Visible learning a synthesis of over 800 meta-analyses relating to achievement*. Routledge.

Ifenthaler, D., & Yau, J. Y.-K. (2020). Utilising learning analytics to support study success in higher education: A systematic review. *Educational Technology Research and Development, 68*(4), 1961–1990. https://doi.org/10.1007/s11423-020-09788-z

Janssen, J., Kirschner, F., & Kirschner, P. A. (2021). The collaboration principle in multimedia learning. In R. E. Mayer & L. Fiorella (Eds.), *The cambridge handbook of multimedia learning* (3 ed., pp. 304-312). Cambridge University Press. https://doi.org/10.1017/9781108894333.032

Johnson, D. W., Johnson, R. T., & Holubec, E. J. (2008). *Cooperation in the classroom* (8th ed.). Interaction Book Co..

Kimmons, R., Graham, C. R., & West, R. E. (2020). The PICRAT model for technology integration in teacher preparation. *Contemporary Issues in Technology and Teacher Education, 20*(1), 176–198.

Kirschner, F., Paas, F., & Kirschner, P. A. (2011). Task complexity as a driver for collaborative learning efficiency: The collective working-memory effect. *Applied Cognitive Psychology, 25*(4), 615–624. https://doi.org/10.1002/acp.1730

Kollar, I., Fischer, F., & Hesse, F. W. (2006). Collaboration scripts – A conceptual analysis. *Educational Psychology Review, 18*(2), 159–185. https://doi.org/10.1007/s10648-006-9007-2

Mayer, R. E. (2002). Multimedia learning. In *Psychology of Learning and Motivation* (Vol. 41, pp. 85–139). Academic Press. https://doi.org/10.1016/S0079-7421(02)80005-6

Mayer, R. E. (2020). *Multimedia learning* (3 ed.). Cambridge University Press. https://doi.org/10.1017/9781316941355

Mayer, R. E. (2021). Evidence-based principles for how to design effective instructional videos. *Journal of Applied Research in Memory and Cognition, 10*(2), 229–240. https://doi.org/10.1016/j.jarmac.2021.03.007

Noetel, M., Griffith, S., Delaney, O., Sanders, T., Parker, P., del Pozo Cruz, B., & Lonsdale, C. (2021). Video improves learning in higher education: A systematic review. *Review of Educational Research, 91*(2), 204–236. https://doi.org/10.3102/0034654321990713

Ramsden, P. (1991). A performance indicator of teaching quality in higher ecucation: The course experience questionnaire. *Studies in Higher Education, 16*(129), 129–149.

Salmon, G. (2013). *E-tivities: The key to active online learning* (2. ed.). Routledge.

Vogel, F., Wecker, C., Kollar, I., & Fischer, F. (2017). Socio-cognitive scaffolding with computer-supported collaboration scripts: A meta-analysis. *Educational Psychology Review, 29*(3), 477–511. https://doi.org/10.1007/s10648-016-9361-7

Chapter 9
Reflection

As we have repeatedly noted, the learning effectiveness of the listed guidelines and principles is highly dependent on contextual conditions. This makes it all the more important to revisit the objectives and contextual conditions at the end of the educational design process to anticipate the learner's journey and learning experience and to reflect on the educational design.

This chapter explores the following key questions: How can the design developed for the digital learning environment be reviewed? How can we ensure that the key steps and conditions for creating effective digital learning environments have been met?

9.1 Review of the Educational Design

The educational design should be reviewed to ensure that it aligns with both the learner needs and the strategic goals of the educational institution. One aspect to evaluate is the design's temporal and spatial flexibility, which can be assessed by analysing the proportion of asynchronous or online learning within the workload. However, determining the ideal proportion is challenging, as it depends on factors such as desired learning outcomes, learner needs, teaching culture, and available infrastructure. The design should also cater to learner diversity and allow adaptability (e.g., through hybrid formats). Additionally, consider whether synchronous and on-site phases are required for educational or organizational reasons.

The concepts of constructive alignment and the iron triangle (discussed in Chap. 3) provide effective frameworks for reviewing educational designs concerning learning effectiveness, efficiency, and attractiveness.

Constructive alignment ensures consistency between objectives, the learning environment, and assessments. This involves confirming that the designed activities support the intended competencies and are assessed appropriately. Mapping out which learning activities align with each objective can offer an early review of how

C. Müller, *Digital Learning Design*, SpringerBriefs in Education,
https://doi.org/10.1007/978-3-031-89045-1_9

well the design supports all goals and whether the learning environment is effective. Following implementation, *learning effectiveness* can be evaluated by outcomes, ensuring that complex competencies are assessed using appropriate methods rather than basic knowledge tests.

The *iron triangle* framework broadens the focus to include efficiency and attractiveness alongside learning effectiveness. A workload analysis can help determine whether learning outcomes are achievable within a given time, which is essential for institutional education (e.g., universities with fixed workloads) and professional training, where *efficiency* is critical.

Attractiveness can be assessed using the ARCS model by addressing key questions (see Sect. 2.2): Does the learning experience capture attention? Does it relate to the learner's world and establish relevance? Does it foster confidence and satisfaction? Indicators of quality include the balance between activating, interactive, and assessment activities versus content delivery, as well as the ratio of active/constructive/interactive tasks to passive ones. The quantity of formative and summative assessments reflects the extent of feedback provided to learners on their progress. Similarly, the number of peer activities indicates opportunities for social interaction.

After initial implementation, *learners' feedback* becomes a valuable resource for identifying areas for improvement, particularly regarding the learning environment's attractiveness. Many educational institutions have developed instruments for evaluating digital learning environments. Baldwin et al. (2018) synthesized these instruments, offering a comprehensive resource to help create tailored evaluation systems suitable for your context.

9.2   Review of the Design Process

After completing the design process, use the checklist provided in Table 9.1 to ensure that all essential steps and conditions for creating effective digital learning environments have been addressed.

Table 9.1 Digital learning design checklist

	Yes	No
Analysis		
• Have the learning needs and required competencies (knowledge and skills) been analysed and documented as part of a task and content analysis?	☐	☐
• Has the target audience been analysed and described in detail?	☐	☐
• Are the learning objectives aligned with target competencies and the learning context and formulated in behavioural observable terms?	☐	☐
Content structuring		
• Is the learning content appropriately prioritized, educationally adapted, segmented and sequenced in relation to the learning objectives and the target group?	☐	☐
Learning organization		
• Does the learning organization, with its temporal and spatial sequence of learning phases (physical attendance/online attendance/self-study), meet the needs of learners, teachers, and the learning institution? • Can the desired learning objectives be achieved with the chosen learning organization?	☐	☐
Teaching strategies		
• Have teaching strategies been evaluated and applied appropriately in relation to the educational context?	☐	☐
Learning activities		
• Are the principles of effective learning media design being followed?	☐	☐
• Are there enough learning assignments to activate learners?	☐	☐
• Are the opportunities for social interaction suitable for the context?	☐	☐
• Do learners have sufficient opportunities to check their progress through formative assessment?	☐	☐
• Is the summative assessment aligned with the learning objectives and the learning environment?	☐	☐
• Does the assessment system incentivize learners to engage intensively with the learning objectives, both individually and as a group?	☐	☐
Accessibility		
• Have accessibility standards been considered and met in the development of the digital learning environment?	☐	☐

 **Tips for Evaluation**

- **Testing:** Have the digital learning environment tested by people willing to provide feedback.
- **Use existing instruments:** Adapt established evaluation instruments to fit your own evaluation needs.
- **Plan for revision:** Allocate time and resources to revise the digital learning offering after an initial implementation and review.

*my*Scripting   **Analytics in myScripting**

The "Analytics" view (see below) in myScripting allows you to analyse and reflect on the designed script throughout the design process. The following analyses are available:

- *Workload*: Compares intended and planned workload (in %).
- *Flexible learning*: Compares the proportion of asynchronous vs. synchronous learning (in %).
- *Learning outcomes*: Tracks the number and workload of topics, subtopics, and activities aligned with the learning outcomes.
- *Constructive alignment*: Measures the proportion of learning outcomes supported by learning activities and assessed through formative or summative assessments.
- *Learning activities*: Compares the workload distribution between content delivery and activation/interaction/assessment activities (in %).
- *ICAP learning assignments*: Counts the number of learning assignments based on the ICAP model.
- *Assessment*: Tracks the number of activities with formative or summative assessments.
- *Social inclusion*: Measures the number of peer activities (in small or large groups).

The script can be exported into the selected LMS to build digital learning environments. It is also possible to export the script as a Word document (e.g., a syllabus) or to create an Excel list of media to be produced (e.g., videos).

Reference

Baldwin, S., Ching, Y.-H., & Hsu, Y.-C. (2018). Online course design in higher education: A review of national and statewide evaluation instruments. *TechTrends, 62*(1), 46–57. https://doi.org/10.1007/s11528-017-0215-z

Glossary

Adaptivity Adaptive learning systems evaluate learners' performance and competencies to personalize experiences by adjusting content, learning paths, and difficulty based on their individual needs.

Animation A dynamic visual representation used to explain processes or complex systems.

Artificial Intelligence (AI) A system's ability to correctly interpret external data, to learn from such data, and to use those learnings to perform human-like cognitive functions, such as understanding natural language, problem-solving, and decision-making.

Asynchronous learning Learning that occurs at flexible, self-paced times without simultaneous interaction (see Synchronous learning).

Assessment The process of collecting, evaluating, interpreting, and providing feedback on learners' performance. Assessments can be:

- *Formative*: Process-oriented, focused on guiding and improving learning through feedback.
- *Summative*: Outcome-oriented, focused on evaluating learning results, often with grades, primarily for qualification and selection.

Augmented reality (AR) A computerized extension of reality in which virtual elements are overlaid in the real world using software, apps, and hardware, such as AR headsets.

Backward design An approach in which learning outcomes are identified first, guiding the design of the learning environment and activities to ensure that learners meet those outcomes.

Badge A digital token or certificate representing a specific skill or knowledge (see also gamification).

Blended learning A mix of asynchronous and synchronous learning, often combining face-to-face classroom teaching with online learning.

Branching video An interactive video in which learners make choices that alter the video's progression and outcomes, demonstrating the consequences of their decisions.

C. Müller, *Digital Learning Design*, SpringerBriefs in Education,
https://doi.org/10.1007/978-3-031-89045-1

Breakout rooms Smaller group spaces within online conferencing platforms, allowing participants to engage in discussions separate from the main session.

Community of inquiry (COI) A model in digital learning that highlights three key elements—social presence (learner interaction), cognitive presence (knowledge construction), and teaching presence (instruction and facilitation)—to foster meaningful, collaborative learning experiences.

Competency The combination of skills, abilities, and behaviours necessary to solve real-world problems, typically categorized as professional, methodological, social, or personal competencies.

Constructive alignment Ensures that learning outcomes, the learning environment, and assessments are aligned to support learners in achieving their intended goals.

Curriculum A structured plan outlining learning outcomes, content, processes, and organization for an educational programme or course over a defined period.

Dysfunctional group processes Group dynamics that hinder effective cooperation or learning, such as the free-rider effect.

Enquiry-based learning A student-centred approach in which learners develop their own research questions, pursue them independently, and find solutions.

Fidelity The degree of realism in a learning environment, ranging from low fidelity (e.g., written case studies) to high fidelity (e.g., virtual simulations closely mirroring real contexts).

Formative See Assessment.

Flipped classroom A teaching method in which learners study content independently (e.g., through videos) and class time is used for interactive activities, such as discussions and problem-solving.

Gamification The integration of game-like elements (e.g., badges) into nongame contexts, including educational settings, to enhance engagement.

Guidance The support provided by teachers to help learners navigate and progress in their learning processes.

Hybrid learning A teaching model that combines online and face-to-face instruction, often used interchangeably with blended learning but which currently refers to synchronous teaching sessions offered simultaneously in both formats.

Hyflex model A flexible teaching format in which learners can choose to attend teaching sessions either on-site, online synchronously, or asynchronously through recorded sessions.

Inert knowledge Knowledge acquired in an educational context that learners struggle to apply in practical or real-world situations.

Interleaving A content structuring technique in which different topics or skills are mixed and practiced together in similar but varied contexts, rather than being learned individually in blocks, to improve retention and understanding.

Inverted classroom See flipped classroom.

Instructor-paced course An (online) course with a set start date and deadlines managed by an instructor.

Jigsaw A collaborative teaching strategy in which learners first become experts on a segment of content and then teach it to their peers.

Learning management system (LMS)   A web-based platform that facilitates the delivery of learning content, the organisation of activities, and communication between learners and instructors (e.g., Moodle, Canvas, Ilias, Olat).

Learning analytics   The collection and analysis of data about learners' interactions with the learning environment used to provide real-time feedback and optimize learning.

Learning outcomes   The specific competencies that learners are expected to achieve by the end of a course or learning unit.

Learning activity   Actions taken by learners to meet learning outcomes.

Learning assignments   Tasks that initiate, guide, and support learning activities within a specific context.

Learning organization   Refers to the temporal and spatial structure of a learning programme or offering.

Microlearning   Short and focused learning units typically delivered via mobile devices.

MOOC (massive open online course)   Large-scale online courses open to anyone without admission restrictions.

Online collaborative learning (OCL)   A teaching strategy where learners engage in problem-solving and knowledge creation in online, usually asynchronous, group discussions.

OER (open educational resources)   Free learning and teaching materials made available under open licenses, such as Creative Commons, for teaching and learning.

Peer assessment and self-assessment   Peer assessment involves learners evaluating each other, while self-assessment requires learners to evaluate their own work.

Portfolio   A curated collection of work showcasing a learner's process, development, and achievements.

Problem-based learning (PBL)   A student-centred approach in which learners work in small groups to explore and solve complex real-world problems with the guidance of a tutor.

Project-based learning   Learning that revolves around solving real-world problems, often provided by clients, requiring creative, constructive solutions.

Scaffolding   Support provided to learners to help them bridge the gap between current and desired knowledge or skills, often through prompts, guidance, or additional resources.

Scripting   The process of designing a detailed learning plan, anticipating how learners will engage with the material, and structuring activities accordingly.

Self-paced Course   An (online) course that allows learners to progress at their own speeds without fixed start dates and deadlines (see also Instructor-paced course).

Social form of learning   The way learners interact during the teaching process, typically categorized as individual, pair, small group (3–5 people), large group (6–12 people), or plenary (>12 people).

Spacing   A learning method in which topics are revisited and studied over spaced intervals of time to enhance retention.

Subject matter expert (SME)   An individual with deep knowledge and expertise in a particular subject area who contributes to the development of educational materials.

Summative   See Assessment.

Syllabus   A document for students that outlines the learning outcomes, content structure, learning organization, assessment methods, and resources for a course.

Synchronous learning   Learning activities that occur in real time with live interactions between learners and instructors.

Teaching strategy   A specific sequence and design of learning activities intended to achieve educational goals.

Virtual reality (VR)   A digital, simulated world where learners can immerse themselves and interact using special devices, such as VR glasses.

Webinar   An online presentation, meeting, or workshop enabling real-time interactions (short for "web-based seminar").

Worked example   A step-by-step solution to a task or problem provided as part of the learning process.

Workload   The estimated time required for an average learner to achieve the course objectives or complete a learning activity.